The Periodic Table
A Visual Guide to the Elements

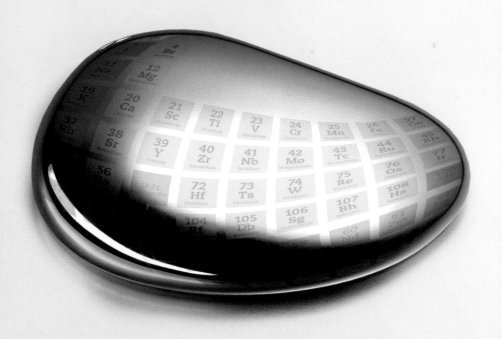

The Periodic Table

A Visual Guide to the Elements

Paul Parsons & Gail Dixon

Contents

The Periodic Table

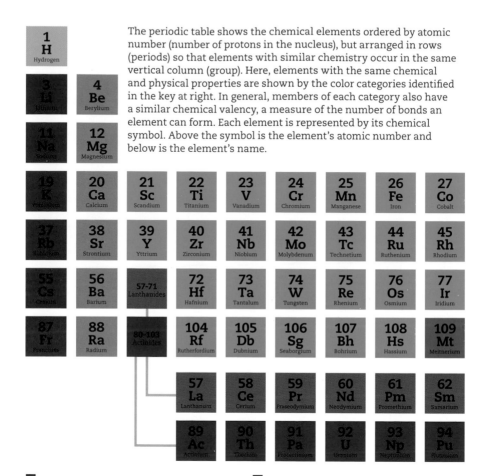

The periodic table shows the chemical elements ordered by atomic number (number of protons in the nucleus), but arranged in rows (periods) so that elements with similar chemistry occur in the same vertical column (group). Here, elements with the same chemical and physical properties are shown by the color categories identified in the key at right. In general, members of each category also have a similar chemical valency, a measure of the number of bonds an element can form. Each element is represented by its chemical symbol. Above the symbol is the element's atomic number and below is the element's name.

Alkali metals
This "group 1" of metals occupies the far-left column of the periodic table. They are all soft, but solid metals at room temperature, and are highly reactive—for example, when dropped in water.

Alkaline earth metals
The alkaline earth metals are silver-white metals at room temperature. The name is a term that refers to the naturally occurring oxides of these elements. For example, lime is the alkaline oxide of calcium.

Lanthanides
The lanthanide elements occupy a horizontal strip normally appended at the foot of the periodic table. Named after lanthanum, the first element in the series, they are generally found in less common mineral rocks, such as monazite and bastnasite.

Actinides
The actinides fill the second horizontal strip at the foot of the table. Named after their first element, actinium, they are all highly radioactive. So much so, that natural reserves of many of these elements have decayed away to nothing.

Transition metals
The transition metals occupy a broad swathe in the center of the periodic table. They are harder than the alkali metals, less reactive, and are generally good conductors of both heat and electrical current.

Post-transition metals
Lying in a triangular region to the right of the transition metals, the post-transition metals are soft metals that mostly have low melting and boiling points. They also include mercury, which is liquid at room temperature.

Element Categories

- Alkali metals
- Alkaline earth metals
- Lanthanides
- Actinides
- Transition metals
- Post-transition metals
- Metalloids
- Other nonmetals
- Halogens
- Noble gases
- Unknown chemical properties

								2 He Helium
5 B Boron	6 C Carbon	7 N Nitrogen	8 O Oxygen	9 F Fluorine	10 Ne Neon			
13 Al Aluminum	14 Si Silicon	15 P Phosphorus	16 S Sulfur	17 Cl Chlorine	18 Ar Argon			
28 Ni Nickel	29 Cu Copper	30 Zn Zinc	31 Ga Gallium	32 Ge Germanium	33 As Arsenic	34 Se Selenium	35 Br Bromine	36 Kr Krypton
46 Pd Palladium	47 Ag Silver	48 Cd Cadmium	49 In Indium	50 Sn Tin	51 Sb Antimony	52 Te Tellurium	53 I Iodine	54 Xe Xenon
78 Pt Platinum	79 Au Gold	80 Hg Mercury	81 Tl Thallium	82 Pb Lead	83 Bi Bismuth	84 Po Polonium	85 At Astatine	86 Rn Radon
110 Ds Darmstadtium	111 Rg Roentgenium	112 Cn Copernicium	113 Uut Ununtrium	114 Fl Flerovium	115 Uup Ununpentium	116 Lv Livermorium	117 Uus Ununseptium	118 Uuo Ununoctium
63 Eu Europium	64 Gd Gadolinium	65 Tb Terbium	66 Dy Dysprosium	67 Ho Holmium	68 Er Erbium	69 Tm Thulium	70 Yb Ytterbium	71 Lu Lutetium
95 Am Americium	96 Cm Curium	97 Bk Berkelium	98 Cf Californium	99 Es Einsteinium	100 Fm Fermium	101 Md Mendelevium	102 No Nobelium	103 Lr Lawrencium

Metalloids

The metalloid elements form a line between the metals and nonmetals in the periodic table. Their electrical conductivity is intermediate between the two groups, leading to their use in semiconductor electronics.

Other nonmetals

In addition to halogens and noble gases, there are other elements that are simply classified as "other nonmetals." They display a wide range of chemical properties and reactivities. They have high ionization energies and electronegativities, and are generally poor conductors of heat and electricity. Most nonmetals have the ability to gain electrons easily. They have lower melting points, boiling points, and densities than the metal elements.

Halogens

The halogens, known as group 17, are the only group to contain all three principal states of matter at room temperature: gas (fluorine and chlorine), liquid (bromine), and solid (iodine and astatine)—all nonmetals.

Noble gases

The noble gases are nonmetals occupying group 18 of the table. They are all gaseous at room temperature and share the properties of being colorless, odorless, and unreactive. Including neon, argon, and xenon, they have applications in lighting and welding.

Unknown chemical properties

This is a label reserved for elements that can only be manufactured in a laboratory. Very often, only minute quantities of such elements have been created—making it impossible to ascertain their exact chemical classification.

Introduction

In the early 1860s, a Russian chemist working at St Petersburg State University came up with an idea that would overturn our understanding of the chemical world. Dmitri Mendeleev put forward the idea of representing all of the known chemical elements in a table, according to their composition and properties. This insight not only enabled him to predict the properties of as-yet undiscovered elements, but would also shape the course of chemical research for evermore.

Mendeleev was fascinated by the chemical elements. Elements are the fundamental building blocks of chemistry—chemical substances that can exist as individual atoms (as opposed to more complex chemical "compounds," the smallest units of which—molecules—are made by joining atoms of different elements together).

Mendeleev wondered if there was any way to order the elements according to their properties. He set about listing all the known elements (there were 62 of them at the time) by what's known as their "atomic mass number." An atom consists of electrically neutral neutron particles and positively charged protons clustered tightly into a central nucleus, around which orbits a cloud of negatively charged electrons. The mass of the electron particles is so small compared to the others that it's usually ignored. The neutrons and protons, however, are heavier; these particles weigh about the same as each other. Counting up the total number of neutrons and protons in an atom's nucleus yields a figure known as the atom's "atomic mass number."

Mendeleev arranged the elements in a long line, left to right, in order of their atomic mass. It was then that he noticed something strange: if he chopped this linear sequence into strips and arranged the strips into rows, to make a table, each column in the table tended to contain elements with similar properties. His left-hand column, for example, contained sodium, lithium, and potassium—all of which are solids at room temperature, tarnish quickly, and react violently with water. Because of this similarity, he referred to the columns of the table as "groups" of elements, while the repetition of properties led him to name the rows "periods." He published his "periodic table" in 1869.

Mendeleev firmly believed that elements should be grouped according to their properties. So much so, he made occasional tweaks to his table, shifting elements on by a column or two to place them in a group with other, more similar, elements. Doing so inevitably left gaps. As it turned out, though, these only served to reinforce the table's validity—by making it testable. Mendeleev asserted that the gaps corresponded to elements that were yet to be discovered. For example, arsenic should have occupied the slot in period 4, group 13. But Mendeleev believed it better fitted with the properties of elements in group 15—which was where he duly placed it. Sure enough, new elements (gallium and germanium) were later discovered that filled the missing slots in groups 13 and 14, and fitted the properties of those groups perfectly.

In places, Mendeleev was so convinced that elements should be grouped by their properties that he even broke the rule of ordering by atomic mass. In so doing, he uncovered the true principle by which the elements are arranged—not by their atomic mass, but by a new property called "atomic number." Whereas atomic mass is the total number of neutrons plus protons in the nucleus, atomic number is determined just by the number of protons it contains. Because protons each carry unit electric charge, atomic number is essentially a measure of the charge on the nucleus, as would be proven in 1913 by the British physicist, Henry Moseley. Atomic number also turns out to be equal to the total number of negatively charged electrons orbiting around the nucleus—making the atom's overall net charge zero. An element can be uniquely specified by its atomic number—for example, carbon is synonymous with "element 6" and plutonium is "element 94."

Moseley's analysis enabled chemists to refine the table further and reveal more gaps, which suggested that there were more new elements waiting to be discovered, with atomic numbers 43, 61, 72, and 75. Scientists subsequently found these elements—now known as technetium, promethium, hafnium, and rhenium, respectively.

Despite Moseley's confirmation of the order of the chemical elements, there was still no real explanation for their periodic properties—why the elements in each group were so similar. It wasn't long, though, before the answer to that question would be discovered via an emerging branch of science known as quantum theory, which gave physicists new insights into the way subatomic particles interact with one another.

When two substances react chemically, what's actually happening is that atoms and molecules of the substances are exchanging and sharing their electrons. Quantum theory determines the behavior of electrons. It revealed that the electrons in an atom are organized into a number of levels, also known as "shells," each of which can accommodate a fixed number of electron particles. As you go from one element to the next, and atomic number increases, each shell gradually fills up—until it's full and the process repeats. Each of Mendeleev's periods corresponds to the filling of an electron shell—each group has the same number of electrons in its outer "valence" shell, which

is the principal determinant of chemical behavior. This is what gives the elements in the same group their similar properties.

Not all electron shells in an atom can hold the same number of electrons. The innermost level, for example, can hold just two; it is this that explains the large gap at the top of the periodic table, where hydrogen (with just one electron in the valence shell) occupies the far left-hand column (group 1) but helium, with just one extra electron, is in the far right column (group 18) among other elements that have a full valence shell. The similar gaps on periods 2 and 3 arise for the same reason. Conversely, the outer shells can hold an inordinately large number of electrons, which accounts for the existence of the lanthanide and actinide sequences—horizontal strips of elements that have been clipped out of the main body of the table and appended at the bottom for neatness.

Although every atom of a given chemical element has a fixed number of protons in its nucleus, the number of neutrons can vary. Atoms with different numbers of neutrons are known as "isotopes"—different isotopes of the same element generally have the same chemistry but different nuclear properties, such as half-life (see Glossary, page 234). Other differences can occur from the way atoms are arranged in a substance. Pure carbon, for example, can take the form of soot, graphite, and diamond. Such different forms of a chemical element are known as "allotropes."

In the text that follows, we meet all the chemical elements that have been discovered so far (there are 118 of them). Each of the first 100 has its own section. The heavier "transfermium" elements (of which there are 18 known beyond element 100) are rarer and have few applications, and so are detailed in a single consolidated section at the back of the book.

Hydrogen

Category: nonmetal
Atomic number: 1

Atomic weight: 1.00794
Color: n/a
Phase: gas

Melting point: -434 °F (-259 °C)
Boiling point: -423 °F (-253 °C)
Crystal structure: n/a

Hydrogen is the number one element of the periodic table and earns this status for a range of reasons: along with helium (see page 14) and lithium (see page 16), it was one of the first three elements produced during the Big Bang; it is the most abundant element in the universe, accounting for 88 percent of all atoms; and it is the lightest of all the elements, with only one proton (which is why it is number 1 in the periodic table) and one electron.

Hydrogen is life-giving in a variety of ways. It is the fuel that keeps our Sun and other stars burning; every time you sunbathe or watch the beauty of a glowing sunset, you are enjoying the result of a massive nuclear reaction. At the Sun's core, the temperature is around 27 million °F (15 million °C) and the density is 250 pounds per pint (200 kg per liter). In such conditions, hydrogen will begin to "burn" in a nuclear process and form helium nuclei, emitting huge amounts of energy.

At standard temperature and pressure, hydrogen is a colorless and odorless gas that exists in the diatomic form H_2 ("diatomic" meaning that it consists of two atoms). In this form, hydrogen is highly combustible and readily forms compounds with other elements. Combined with oxygen (see page 26), hydrogen forms the water that fills the seas, rivers, lakes, and clouds. Allied with carbon (see page 22), it helps to bond the cells of living beings.

Hydrogen is also abundant in the Earth's crust, in the hydrocarbons that have been formed from decaying organisms. These have become our modern-day fuels, such as crude oil and natural gas. Now scientists believe that hydrocarbons may also be formed in the deep Earth, from methane that's subjected to extreme pressures and temperatures.

Hydrogen is a key element in acids and it was this aspect of its chemistry that led to its discovery. In 1766, Henry Cavendish, a wealthy British man with an interest in science, observed bubbles of gas rising from a reaction of iron filings in dilute sulfuric acid. He collected the gas and found it to be highly flammable and very light: qualities that made the gas seem unusual to Cavendish. He was also the first person to prove that when hydrogen burned it formed water, showing that water could be made from another substance and thus disproving Aristotle's theory that there were four basic elements, of which water was one.

Hydrogen is lighter than air and as a result was used in balloon flight, which became popular during the 19th century, and in the airships that crossed Europe on scheduled flights in the early 1900s. During the First World War, Zeppelins were used for reconnaissance missions and bombing raids on London—their light weight kept them out of the reach of fighter planes. The Hindenburg crash of 1937 (when a Zeppelin burst into flames while attempting to land) ended the airship era, although this was not caused by a hydrogen leak, as assumed at the time.

Today, around 55 million tons (50 million tonnes) of hydrogen are produced yearly and large quantities go into fertilizer production. Nitrogen (see page 24) and hydrogen are used as part of the Haber–Bosch process, which

H ¹

This reaction between pieces of zinc metal and hydrochloric acid produces bubbles of hydrogen. Hydrochloric acid molecules are made of hydrogen and chlorine atoms. The zinc is more reactive than hydrogen and replaces it to make soluble zinc chloride. Each hydrogen atom turfed out of the acid combines with another to form diatomic hydrogen gas.

uses natural gas and air to create ammonia—an important raw material in fertilizer production. Fritz Haber won the 1918 Nobel Prize in Chemistry for this discovery and his colleague Carl Bosch won a Nobel Prize in 1931 for the development of high-pressure methods in chemistry.

Hydrogen is the key element in the thermonuclear bomb, which produces sufficient explosive energy, through nuclear fusion between the hydrogen isotopes deuterium and tritium, to wipe out entire cities. Such weapons currently require a nuclear fission explosion (see uranium, page 210) to kick-start the fusion process. Research is now focused on producing a thermonuclear weapon that would not require a fission reaction to activate it. A process called Inertial Confinement Fusion would use a high-energy laser beam to condense hydrogen to a temperature and density that would ignite a fusion reaction.

Hydrogen's presence in water confirms its number one status on the periodic table. It also explains some of the strange chemistry behind water and ice. If you've ever wondered why ice cubes float in water, and vast swathes of ice float on the surface of seas and oceans, the answer lies in hydrogen. Solids are usually denser than liquids because, as most liquids cool, the molecules slow down and get closer together as they eventually form a solid. This tends to make solids denser than liquids, so you'd expect ice to sink in water. However, as water cools down to 39 °F (4 °C) and the molecules slow down, hydrogen bonds occur, which allow one water molecule to link with four other water molecules. This creates an open, crystalline lattice where the molecules are spread out and less dense in a given space. Hence ice is less dense than water and will float on its surface.

There's a similar explanation behind the annoyance of burst pipes in winter. If the water temperature is really low, the latticework of molecules in the ice makes it expand and the resultant buildup of pressure against the sides of the pipe causes it to burst.

This is why it's important to keep your heating on low during winter, even if you're going away, and to insulate pipes that are in attics or close to external walls.

Water vapor is the invisible gaseous state of water and is particularly important for regulating temperature on Earth. As a potent greenhouse gas, water vapor regulates the temperature of Earth so that it is warm enough to support human, animal, and plant life. It is removed from air via condensation, and replaced by evaporation and transpiration (water loss from plants). It's a key aspect of Earth's water cycle and is vital for the formation of clouds and precipitation (rain, sleet, and snow).

Water vapor explains why dew forms on plants and why delicate silk draperies of spiders' webs become noticeable in the morning after a cold, foggy night. As the exposed surface cools down, water vapor condenses at a rate greater than it can evaporate, which results in the formation of water droplets. If temperatures are low enough, dew will become the icy breath of Jack Frost.

This striking mushroom cloud was produced by the detonation of Castle Romeo, an 11-megaton thermonuclear bomb, on March 26 1954. Thermonuclear, or fusion, bombs contain isotopes of hydrogen that are fused together by great heat (which is normally generated by a smaller nuclear "fission" explosion). The fusion process releases vast amounts of energy locked away inside the nuclei of atoms.

Helium

Category: noble gas
Atomic number: 2

Atomic weight: 4.002602
Color: n/a
Phase: gas

Melting point: -458 °F (-272 °C)
Boiling point: -452 °F (-269 °C)
Crystal structure: n/a

Helium is named for the Greek God of the sun, Helios, after it was first detected as unknown yellow spectral lines in sunlight. In India in 1868, French astronomer Jules Janssen passed sunlight through a prism during a solar eclipse (to split the light into its component colors). He found a sudden jump in the brightness of yellow light, which he initially believed to be caused by sodium.

The same year British astronomer Norman Lockyer and chemist Edward Frankland observed the same yellow line in the solar spectrum while sky-gazing over smoky London; Lockyer caused controversy at the time by claiming that it was a new element present in the Sun. He had the last laugh, however, when his hypothesis became accepted within the scientific community. Lockyer also chose the element's name.

Along with hydrogen and lithium, helium was formed during the Big Bang. It is the second most abundant element in the Universe (after hydrogen), making up 23 percent of its matter. On Earth, helium is present in some minerals but most of it is sourced from natural gas.

Helium is a member of the noble gases (group 18 of the periodic table), which are odorless, colorless, have very low chemical reactivity, and have molecules consisting of a single atom. The group gains its "noble" name because the gases do not bond with other "riff-raff" elements; the reason for this is that their outer shell of electrons is "full," giving them little tendency to react chemically.

One of helium's key uses is in cryogenics, most notably in the cooling of superconducting magnets in MRI scanners. Liquid helium at a temperature of -452 °F (-269 °C) cools the magnets used, to create the intense magnetic field required. Radio waves within the magnetic field interact with hydrogen atoms in water and other molecules in the body in order to produce an image that enables doctors to identify tumors and other medical problems.

Being lighter than air, helium is used in airships and balloons. Helium has also been used as a medium to force rocket fuel and oxidizers out of their storage tanks and to actually make rocket fuel by using the low temperature of liquid helium to condense hydrogen and oxygen into liquid form. It is also used to purge fuel from spacecraft launch gantries to prevent them from igniting dangerously during liftoff. The Saturn V booster used in the Apollo program required 13 million cubic feet) (400,000 cu m) of helium to launch.

One of helium's isotopes, helium-3, has sparked considerable interest as a potentially safe, environmentally friendly fuel. While it's scarce on Earth, it's plentiful on the Moon, which adds weight to the argument for a NASA Moon base. One of the space agency's goals is to mine the Moon for fuel to be used in fusion reactors—futuristic power plants that would produce no carbon dioxide, thus reducing damage to the environment—and Helium-3 is considered a possible fuel for such reactors.

He 2

A gas-discharge lamp filled with helium gives off a ghostly purple glow. Electric current ionizes the gas, tearing electrons from their host atoms and permitting a current to flow through the gas, causing it to glow.

Lithium

Category: alkali metal
Atomic number: 3

Atomic weight: 6.941
Color: silver–white
Phase: solid

Melting point: 358 °F (181 °C)
Boiling point: 2,448 °F (1,342 °C)
Crystal structure: body-centered cubic

Along with hydrogen and helium, lithium is one of the three elements to be created during the Big Bang. Under standard conditions, this soft, silvery white alkali metal is the lightest of all metals and the least dense solid element. Lithium is highly reactive and has to be stored with a coating of petroleum jelly.

Lithium carbonate has been life-changing for thousands of people across the world. In the 1940s, electroconvulsive therapy and lobotomy were considered "standard" treatments for people with mental health issues. Such extremes were to come to an end for many in 1949 following the work of Australian doctor John Cade, a senior medical officer in the Victoria Department of Mental Hygiene, who discovered that when guinea pigs were given lithium carbonate, these normally highly strung animals became so placid that they would lie on their backs contentedly for hours. Cade experimented by taking lithium himself before giving it to one of his patients, who had been in the secure unit for many years. Within a few days the patient was transferred to a standard ward and a few months later he returned home and took up his old job.

Other patients with mental health issues responded well to lithium carbonate, but toxicity was a problem in the early days. More than 10 milligrams of lithium per liter of blood can lead to mild lithium poisoning and levels above 20 milligrams per liter can be fatal. Dosage levels are adjusted according to patients' needs so that the level of lithium in blood plasma is between 3.5 and 8 milligrams per liter. Within 10 years of Cade's findings, lithium was widely used to treat patients across the world. Little is understood of how lithium carbonate can help to smooth out the highs and lows that patients with bipolar disorder experience; it is thought, however, to reduce levels of a chemical messenger that can be overproduced in the brain.

Intriguingly, the therapeutic powers of lithium may have been "discovered" as early as the second century, when the physician Soranus of Ephesus treated patients with "mania" and "melancholia." He used the alkaline waters of his town, which we know contained high levels of lithium. The metal itself was first identified by Johan August Arfvedson, who detected the presence of a new element while working on petalite ore in 1817.

Lithium's main ores today are spodumene, petalite, and lepidolite. Large deposits of spodumene have been found in South Dakota in the United States of America (USA), as well as in Russia, China, Zimbabwe, and Brazil. Lithium is also extracted from the brines of lakes, most notably those in Chile, and California and Nevada in the USA.

Despite its reactivity, lithium is useful in many everyday items. Lithium oxide is used in heat-resistant glass and ovenware, while lithium-ion batteries pack considerable power into small spaces and so are ideal for pacemakers, watches, laptops, and cameras. Lithium is also important in the construction of airplanes—it combines with aluminum to form an alloy that's lighter than pure aluminum, thus reducing fuel consumption.

Samples of the alkali metal lithium, which is soft
enough at room temperature to be cut with a knife.
Lithium has applications in areas ranging from
battery technology to mental health medication.

Beryllium

Category: alkaline earth metal
Atomic number: 4

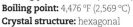

Atomic weight: 9.012182
Color: silvery white
Phase: solid

Melting point: 2,349 °F (1,287 °C)
Boiling point: 4,476 °F (2,569 °C)
Crystal structure: hexagonal

Beryllium is the fourth lightest element, but it is not very abundant in the Universe, because it is only generated during supernova explosions. The element is named after *beryllo*, a Greek word meaning beryl, a mineral ore that contains beryllium (bertrandite is another). Sometimes huge crystals of the mineral are discovered, some up to 20 feet (6 m) long, and the key areas for ores are in the USA, Brazil, Russia, India, and Madagascar. One of its isotopes, radioactive beryllium-10, is generated when cosmic rays react with oxygen and this has been detected in ice cores in Greenland.

The Romans and ancient Egyptians prized emeralds and beryl, and the Roman writer and philosopher Pliny the Elder, writing in the first century AD, observed that they were both derived from the same mineral. It was not until the late 18th century that chemists in France noticed a previously unknown element in beryl ore, and the new element beryllium was successfully isolated in 1828.

Beryllium is a lightweight, silvery white metal. It has no known biological role and it is toxic to humans. If beryllium fumes are breathed in, they cause a persistent lung condition called berylliosis, which includes symptoms such as inflammation of the lungs and shortness of breath. Workers exposed to beryllium alloys in the mid-20th century were most at risk, when early fluorescent lamps were coated with phosphors containing beryllium oxide. Manufacture of this lamp was ended when large numbers of workers at a plant in the USA contracted the lung disease.

Beryllium has been key in the development of atomic theory because of the role it played in the discovery of the neutron. In the early 20th century, physicists believed that—due to measurements of atomic mass—nuclei must contain particles other than positively charged protons. British physicist Sir James Chadwick spent a decade exploring this field of atomic theory and, in 1932, discovered that if beryllium was bombarded with alpha rays from radium, it would emit an unknown subatomic particle. This particle had about the same mass as a proton but no electrical charge. Chadwick had discovered the neutron and, in 1935, received the Nobel Prize in Physics for his discovery.

If a few percent of beryllium is alloyed with copper, it forms a nonsparking, high-strength metal that is ideal for tools used in environments such as oil wells or where there are flammable substances and just one spark could spell disaster. Beryllium is a strong metal, which is lightweight, resistant to corrosion, and melts at a very high temperature. All these factors combined make it ideal for use in spacecraft and missile components. Beryllium is also used in the windows of X-ray tubes, because it is transparent to X-rays.

The element is unusual in its ability to reflect neutrons, which is one reason why it is used in nuclear weapons. Inside the warhead, neutrons bombard uranium in order to release energy. As the neutrons are reflected back and forth contained by the beryllium casing, this increases the rate of the nuclear reaction within the weapon.

Be ^{4}

The alkaline earth metal, beryllium. Although X-rays stream through it as if it were made of glass, this metal is highly reflective to neutron particles (leading to its use in nuclear weapons).

Boron

Category: metalloid
Atomic number: 5

Atomic weight: 10.8111
Color: varied
Phase: solid

Melting point: 3,769 °F (2,076 °C)
Boiling point: 7,101 °F (3,927 °C)
Crystal structure: rhombohedral (as boride)

Boron is one of the unsung heroes of the periodic table. It may lack the glamor of gold or platinum, and have the dullest of names, but it can conjure intriguing chemical wizardry and its compounds have proved useful for centuries.

Boron is a key element in borax, also known as sodium borate, sodium tetraborate, or disodium tetraborate. This is a salt of boric acid, which usually occurs as a white powder and dissolves easily in water. It has been used as a detergent, fungicide, and pesticide for millennia and also as a fire retardant, a component in ceramics, and as a flux to make molten metal easier to work.

The name boron is derived from the Arabic *buraq*, meaning "borax." The salt was first discovered in Tibet and transported west via the Silk Road (an ancient trade route that linked China with the West) to the Near East and Europe. Borax even gets a mention in "The Prologue" of the English poet Geoffrey Chaucer's *Canterbury Tales*, written in the late 14th century. Borax was later used as a cosmetic and was mixed with oil, powdered egg shell, and other ingredients to create the virginal-white look favored by England's Queen Elizabeth I.

The presence of boron itself was first noticed in 1732 as an unusual green flame by the French chemist, Geoffroy the Younger. This can be achieved by treating borax with sulfuric acid to change the boron into boric acid. Alcohol is then added and the acid is set alight. In 1808, British chemist Sir Humphry Davy tried to isolate pure boron from borate by heating it with potassium metal. In Paris,

Joseph Louis Gay-Lussac and Louis-Jacques Thénard were following the same procedure, but neither they nor Sir Humphry could isolate the pure element. It wasn't until 1892 that Ezekiel Weintraub of the American General Electrical Company, sparked a mix of boron chloride vapor and hydrogen.

Boron is only found as borate minerals on Earth. Large deposits of borates have been located in California's Death Valley, in the USA, and also in Tuscany, Italy. Today, it is mined mostly in the USA, Turkey, Tibet, and Chile.

Several allotropes (different forms) of boron exist. The most common is amorphous boron, a dark powder that reacts with metals to form borides but does not react with oxygen, water, or acids and alkalis.

Intriguingly, if boron is combined with nitrogen, the crystals produced (boron nitride) are almost as hard as diamond. These are much cheaper to create and have greater heat resistance, which makes them highly useful abrasives in steelworking. (In industry, abrasives are used for cleaning, removal of excess material, sharpening, and cutting.)

If you have some Silly Putty™ (also known as Thinking Putty™) on your desk, you might find it interesting to know that boron cross-links the polymers that give these colorful, silicone-based splodges their ability to be both soft and malleable in the hand but bouncy and hard when chucked around. Try it: it might help when considering the magnitude of the next element of the periodic table—the behemoth that is carbon.

A sample of pure, crystalline boron. You won't find anything like this in nature, though, as boron only exists naturally in the form of compounds with other elements—such as the minerals kernite and colmanite.

Carbon

Category: nonmetal
Atomic number: 6

Atomic weight: 12.0107
Color: clear (diamond) and black (graphite)
Phase: solid

Melting point: n/a—turns to vapor (sublimes) before it melts
Sublimation point: 6,588°F (3,642°C)
Crystal structure: hexagonal (graphite) and face-centered cubic (diamond)

Carbon is one of the most important elements in the periodic table—at least for us here on Earth; its rich spectrum of chemical compounds make it the backbone of life on our planet. Carbon helps to power the nuclear reactions in the Sun that provide us with light and warmth, and it is also a key component of our technology—integral to the fuel that powers civilization, the materials in our buildings, and even the clothes on our backs.

Carbon is an element from antiquity, known to the ancient Egyptians and Sumerians as far back as 3750 BC, who used it for processing ores in the manufacture of bronze. Later, it would find another application in metallurgy, as the element used to convert raw iron into a superstrong alloy: steel. The name "carbon" derives from the Latin *carbo*, meaning "charcoal."

Several different forms of carbon were known to the ancients—such as charcoal, coal, and graphite—but it wasn't clear to them that these substances were just different facets of the same thing. This was established chemically in the 18th century. In 1779, German–Swedish chemist Carl Wilhelm Scheele showed that graphite contained carbon. Seven years later, a French team showed that graphite was, in fact, made mostly of carbon.

Carbon is the 15th most abundant element in the Earth's crust; the planet's biosphere alone is home to about 2,000 billion tons of the stuff. Carbon is a tetravalent atom, able to bond with up to four other atoms simultaneously, enabling it to form over 20 million different chemical compounds—this diversity is central to carbon's role at the heart of life on Earth. This is described in part by the "carbon cycle," which governs how carbon is taken in by plants (through photosynthesis) from carbon dioxide in the atmosphere and then ingested by animals, before being returned to the atmosphere when these life forms ultimately die.

Current environmental concerns hinge on the fact that a great deal of carbon is being added to the cycle by the digging up and burning of fossil fuels. The carbon in these substances—such as coal and oil—is turned into carbon dioxide when burned. In the atmosphere, this gas acts to trap heat from the Sun, warming the planet.

Carbon has many uses. As well as being used in steelmaking, it is the basis of plastics, cotton, and linen. Carbon (in its form as charcoal) was also a key ingredient in the first practical explosive: gunpowder.

Carbon comes in various different allotropes, which are essentially different substances formed by arranging the carbon atoms in different ways. Graphite is an allotrope made from sheets of bonded carbon atoms. The gemstone diamond, however, has a rigid interlocking lattice of atoms, formed at a high temperature and pressure. Most recently, an extraordinary form of carbon known as a fullerene has been discovered. Fullerenes consist of graphite rolled into cylinders (nanotubes) or spheres (known as buckyballs). They are hundreds of times stronger than steel.

Two forms of carbon—a graphite rod and industrial diamonds. These are known as "allotropes" of carbon, the same atoms just arranged in different ways.

Nitrogen

Category: nonmetal
Atomic number: 7

Atomic weight: 14.00672
Color: none
Phase: gas

Melting point: -346 °F (-210 °C)
Boiling point: -320 °F (-196 °C)
Crystal structure: n/a

It's incredible to think that 78 percent of our atmosphere is made of the element nitrogen. Earth's abundance of nitrogen was caused by volcanic outgassing early in its history; by comparison, the element makes up just 2.6 percent of the atmosphere on the planet Mars.

Nitrogen was discovered in 1772, by the British chemist Daniel Rutherford. In the 1760s, scientists Henry Cavendish and Joseph Priestley carried out experiments in which they removed all the oxygen from air. Quite what was left no one knew. It was their compatriot Rutherford who realized that the gas was a new element—subsequently named nitrogen. The name comes from the Greek words *nitron* and *genes*, together meaning "niter-forming"—niter was then the name for saltpeter, or potassium nitrate, the major component of gunpowder.

Nitrogen-based compounds crop up frequently in the history of explosives. That's because they tend to want to release their nitrogen back into gas form extremely rapidly—along with a large quantity of heat. Nitroglycerin is made by reacting nitric acid with glycerin to produce an explosive liquid detonated by impacts—invented in 1846, it was the first "high explosive." Alfred Nobel invented dynamite in 1866 by absorbing nitroglycerin into kieselguhr—a soft powdered sedimentary rock—to create a much more stable explosive. The fortune he made from dynamite funds the prizes that now bear his name.

Ironically, one nitrogen-based explosive, called azide, has saved many lives through its use in car airbags. An accelerometer detects the impact and detonates the charge with an electrical impulse—filling the bag with gas in 25 thousandths of a second. It is estimated that between 1990 and 2000, airbags saved more than 6,000 lives in the USA alone.

Nitrogen is also a useful "inert" atmosphere that prevents damage caused by oxygen. For example, a piece of fruit stored in a sealed (but nonrefrigerated) box of nitrogen will keep for up to two years. Liquid nitrogen is an effective coolant, often used for freezing blood and transplant organs. It also forms the basis of "hydrazine" rocket fuel and "widgets" in canned beers, which release a surge of nitrogen gas to "froth up" the beer. The human body is three percent nitrogen by mass, incorporating the element into proteins, neurotransmitters, and deoxyribonucleic acid (DNA). The compound nitrous oxide (N_2O, "laughing gas") is the gas in the anesthetic "gas and air," offered to pregnant women during labor.

Nitrogen is also an essential nutrient for plants, and most plants take it in through their roots rather than taking it directly from the air. Nitrogen from dead organisms enters the soil, much of which is then consumed and released as gas by bacteria.

Recognizing plants' need for nitrogen has been important for agriculture. In 1909, German chemist Fritz Haber came up with an efficient way to make ammonium nitrate fertilizer to enrich land on which crops are grown. In the last 20 years, these chemicals have doubled crop yields worldwide and are directly responsible for feeding a third of the planet's population.

At some restaurants, diners can eat food that's been poached in liquid nitrogen before their eyes. Nitrogen only exists in its liquid state at temperatures below -320 °F (-196 °C).

Oxygen

Category: nonmetal
Atomic number: 8

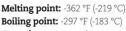

Atomic weight: 15.9994
Color: colorless
Phase: gas

Melting point: -362 °F (-219 °C)
Boiling point: -297 °F (-183 °C)
Crystal structure: n/a

Oxygen has been one of the most essential elements in mankind's evolution. It is the most abundant element on Earth, constituting half of the mass of the planet's crust and 86 percent of the mass of the oceans. It is also the third most abundant element in the Universe, after hydrogen and helium, although our planet is unusual in the Solar System for having such a high level of oxygen gas. Oxygen is highly reactive and will form oxides with most other elements, apart from helium and some of the inert gases. At standard temperature and pressure, two oxygen atoms combine to form O_2, the colorless, odorless gas that makes up around 21 percent of air and is vital for almost all forms of life on Earth.

To understand fully oxygen's life-giving role on Earth, we have to journey back three billion years, when changes in oxygen levels triggered the move toward biodiversity. It was at this point that blue-green algae started to develop and photosynthesis began. This is the process by which plants use carbon dioxide and light in order to produce energy; oxygen is released into the atmosphere as a result.

For around a billion years, most of the extra oxygen produced went into the oxidation of iron. Around 500 million years ago, as land plants began to grow, atmospheric oxygen reached the level of around 21 percent volume, where it has remained ever since. That rise in atmospheric oxygen enabled biodiversity because multicellular organisms, and higher life forms such as human beings and animals, require oxygen to thrive. Thus oxygen not only sustains human life on Earth, it also helped to create us in the first place.

Oxygen's discovery can be attributed to three 18th-century chemists (although around 300 years earlier, Leonardo Da Vinci noted that air contained something vital for human and animal life and that without it a candle flame could not burn). In 1772, Swedish chemist Carl Wilhelm Scheele (see also carbon, page 22) became the first scientist to isolate oxygen by heating mercuric oxide and a number of nitrates; however, his discovery was not published until five years later. In the meantime, British chemist Joseph Priestley had produced oxygen by thermal decomposition of mercuric oxide and published his findings in 1774. He informed the chemist Antoine Lavoisier of his discovery, and the Frenchman became the first scientist to grasp the great importance of oxygen in chemical processes on Earth: respiration and combustion. As a result, scientific theories that had stood for a hundred years were overturned. Lavoisier believed (mistakenly) that all acids contained oxygen, so he named the element oxygène —oxy meaning "sharp" and gen meaning "producing." Although erroneous, the name stuck and passed into common usage.

Oxygen is required by almost every cell in the body for a number of reasons, including metabolism, growth of new tissue, and disposal of waste matter. If the brain is deprived of oxygen for a few minutes, its cells will start to die. After we inhale,

O 8

Oxygen in water: the colder the water is, the more dissolved oxygen it can hold. This is why you will sometimes see bubbles forming on the inside of a glass of tap water: as the water warms up some of the oxygen it holds is forced out of solution.

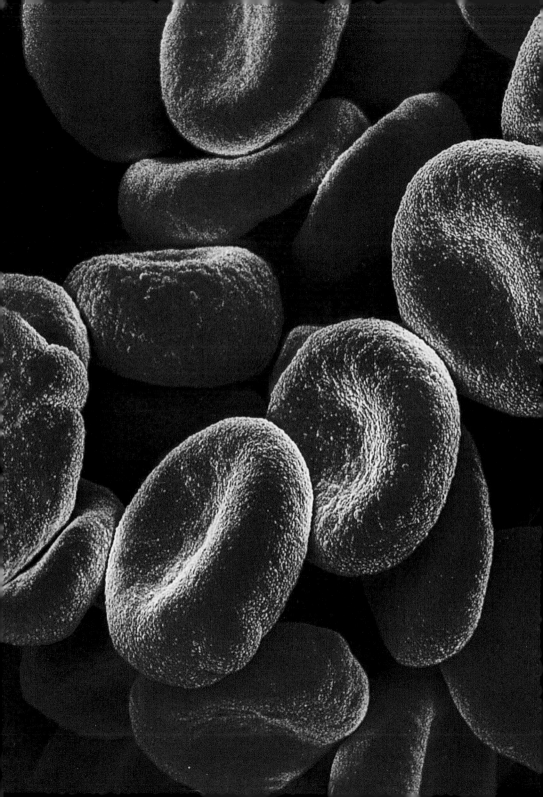

oxygen in the lungs is picked up by the iron atom in hemoglobin (a protein found in red blood cells) and carried around the body to organs and tissue. Oxygen is a component of DNA and almost every other biological compound. It comprises 60 percent of the mass of human beings, because so much of our body is made up of water (which is a hydrogen–oxygen compound). It's in our bones and teeth and in our blood, sweat, and tears.

While being essential for life on Earth, oxygen is also responsible for forms of decay, such as food spoilage. Oxygen is tasteless, colorless, and has no smell, so it's easy to overlook it as a factor; however, the oxygen in air provides ideal conditions for the growth of microorganisms, molds, and yeast, which feed and grow on the surface of foods. Also, oxidizing enzymes found in food speed up the chemical reactions between oxygen and that food, making it "go off." Oxidation itself can also cause food to spoil because when fats oxidize, compounds are formed that produce a very strong, rancid odor. Mankind has developed many resourceful forms of preservation, from burying food in the ground—to keep it at a low temperature and help protect it from air—to drying, curing, canning, bottling, freezing, and vacuum packing.

The percentage of oxygen in the air is constant up to an altitude of about 13 miles (21 km); however, as altitude increases, air density decreases, which means that there are fewer oxygen atoms the higher you go. This results in altitude sickness, the symptoms of which include dizziness, nausea, headache, and exhaustion. The condition can be fatal, due to fluid buildup on the lungs or the brain. Mountaineers and skiers who spend time at high altitudes are careful to acclimatize to their surroundings, ascending the tallest peaks slowly and increasing altitude by no more than 1,000 feet (300 meters) per day. Most cases of altitude sickness occur at heights of 11,500 feet (3,500 meters) and above. The summit of the highest mountain in North America, Mount McKinley (also known as Denali), in Alaska is at 20,322 feet (6,194 meters) above sea level. Mount Everest, in Nepal, is 29,029 feet (8,848 meters) above sea level.

Apart from O_2, the other main allotrope of oxygen is ozone, which consists of three oxygen atoms bonded together. In the stratosphere, 12–25 miles (20–40 km) above Earth, the ozone layer provides a protective shield for Earth from the Sun's damaging ultraviolet rays. Some chemical pollutants destroy ozone, and so weaken the protective shield it provides—these pollutants include emissions from vehicle exhausts, industrial facilities, and fossil-fuel power stations. In the lower part of the atmosphere, however, ozone itself is a pollutant, formed by chemical reactions between oxides of nitrogen and volatile organic compounds in sunlight. Breathing ozone can cause chest pain, coughing, and congestion. It can also worsen asthma and bronchitis.

Oxygen gives us life, but we have used it in such a manner that it could make life on Earth highly dangerous for mankind. Its reactivity has been essential in the process of combustion, fueling factories and motor vehicles, and thus increasing carbon dioxide emissions and global warming. It took billions of years for oxygen levels to rise to enable biodiversity and the birth of mankind. It has taken just 300 years of industrialization to trigger a more dangerous turn in the planet's ecological course.

Red blood cells are used by the body to transport oxygen from your lungs, where it is absorbed from the air, to your organs, where it gets put to work in biochemical reactions that generate energy. Disgraced cyclist Lance Armstrong, and other dishonest endurance athletes, are believed to have taken a substance called "erythropoetin" which boosts red blood cell production, enhancing the oxygen-carrying capacity of the blood and making more energy available to the athletes' muscles.

Fluorine

Category: halogen
Atomic number: 9

Atomic weight: 18.9984032
Color: pale yellow
Phase: gas

Melting point: -364 °F (-220 °C)
Boiling point: -307 °F (-188 °C)
Crystal structure: n/a

Fluorine is one of the most reactive and dangerous of the elements. Exposure to this pungent gas at just 0.1 percent of air for a few minutes can be lethal, and if a stream of fluorine is directed at almost anything—even a pane of glass or a brick—it will erupt into flames. Fluorine is a member of the halogen group of the periodic table, along with chlorine, bromine, iodine, and astatine, and all share the characteristic of being highly reactive.

In the 1520s, it was noted that the mineral fluorspar (calcium fluoride) would melt and flow when heated in a fire (fluorine's name comes from the Latin *fluere*, meaning "to flow"). This led to its use as a "flux"; when added to furnaces, it would make the metallic smelt more fluid and therefore easier to work with.

Alchemists realized that fluorides contained an as-yet unidentified element, which was similar to chlorine. Isolating it proved a huge risk, however, due to its toxicity; a number of chemists died trying to do so and were dubbed "the fluorine martyrs." In 1886, French chemist Henri Moissan was able to obtain fluorine through electrolysis; he was awarded the Nobel Prize in Chemistry in 1906 for this work.

Fluorine may be reactive and dangerous in its natural state but it is stable as fluoride, the form in which it is encountered naturally. Fluoride is considered essential for humans, although it is required in very small amounts—approximately 3 milligrams (0.0001 oz) a day is sufficient. In 1945, water fluoridation began in some areas of the USA, after scientists noted that people who lived in areas with higher water fluoride levels had fewer dental cavities. The fluoridation of water is standard practice in many countries.

Levels of fluoride have to be closely monitored, however, because exposure to high levels can lead to skeletal fluorosis, where fluoride accumulates in the bones and leads to deformities. The US Environmental Protection Agency has set a maximum amount of fluoride in drinking water of 4 milligrams per liter. Fluoride is added to toothpaste because when it is applied to teeth it forms crystals that make enamel more resistant to acid, which can ordinarily make the tooth surface porous and prone to decay.

The most resilient fluorine compound has become a byword in modern times: Teflon™, or polytetrafluororethene, if you want to know its chemical name. Teflon™ was discovered in 1938 by Roy Plunkett at DuPont™ research laboratories, New Jersey, USA, who was researching new chlorofluorocarbon refrigerants. Plunkett had stored polytetrafluoroethylene gas in small cylinders for chlorination and on opening one of these cylinders, found a white powder, which he decided to test. The substance was a plastic that was heat resistant, chemically inert, and remained flexible down to -400 °F (-240 °C), which made it useful for use on space flights. Another property of the compound was that nothing would stick to it, which led to its use as a coating to create nonstick pans. Teflon™ is also used in breathable clothing, which allows water vapor out but resists liquid water getting in.

F ⁹

A fluorite crystal—a kind of mineral made of the compound calcium fluoride. Fluorite samples are often "fluorescent," meaning that they give off light—this mineral is the origin of the word.

Neon

Category: noble gas
Atomic number: 10

Atomic weight: 20.1797
Color: colorless
Phase: gas

Melting point: -415 °F (-249 °C)
Boiling point: -411 °F (-246 °C)
Crystal structure: n/a

What would the Las Vegas Strip be without the glitz, glamor, or indeed gaudiness supplied by neon light? What would Piccadilly Circus in London, or Times Square in New York, look like after dark without their glowing beacons and signage? Neon may be a colorless gas under standard conditions, but if high-voltage electricity is run through a tube filled with neon, it will create a dazzling red–orange glow.

Neon is a member of group 18 of the periodic table, which includes the noble (or inert) gases—helium (page 14), argon (page 50), krypton (page 88), xenon (page 128), radon (page 198), and element-118 (page 232). The name "noble" refers to the fact that most of the members in the group do not react with other elements. In fact, neon is the least reactive element of the group and of the periodic table itself. It will not form compounds with any other elements and has no biological role. Neon is one of the most abundant elements in the Universe, but it occurs naturally in Earth's atmosphere only in trace amounts (around one part in 65,000).

Neon was discovered in 1898 by the British chemists Sir William Ramsay and Morris W. Travers, using spectroscopy (analysing the light produced when an element is heated). Ramsay and Travers chilled air until it became liquid, then warmed it and captured the gases as they evaporated. After nitrogen, oxygen, and argon were removed, they examined the tiny amount of gas remaining. On heating it, they saw astonishing spectral lines and said of the discovery: "The blaze of crimson light from the tube told its own story . . . Nothing in the world gave a glow such as we had seen." Neon produces only a reddish orange glow and no other color.

Ramsay's son, Willie, wanted to name the discovery *novum*, from the Latin word meaning "new." His father liked the idea but preferred the Greek word for "new," *neos*— hence the gas became known as neon.

Today, the element is used mostly in the eponymous lights that capture our attention throughout the world. The first neon tube lights were produced in 1910 by the French engineer Georges Claude and they rapidly gained popularity in advertising due to their arresting quality.

Neon is also used in vacuum tubes, television tubes, lasers, and high-voltage switching gear. Liquefied neon can maintain a temperature as low as -411 °F (-246 °C) and as a result is used as an efficient refrigerant for very low temperatures.

The element is not only found in the atmosphere. Traces of neon have been detected in volcanic fumaroles (openings in the Earth's crust that emit steam and gases). In 1909, the eminent French chemist Armand Gautier detected neon in gases that were emitted from fumaroles near Vesuvius (the active volcano near Naples in southern Italy) and from thermal springs in the same region.

Neon gas in a borosilicate vial. Silver-coated copper wire around the outside passes an electric current through the gas, causing it to glow—this is the principle on which neon lights operate.

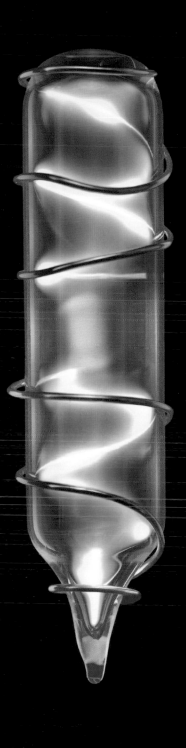

Sodium

Category: alkali metal
Atomic number: 11

Atomic weight: 22.989770
Color: silvery white
Phase: solid

Melting point: 208 °F (98 °C)
Boiling point: 1,621 °F (883 °C)
Crystal structure: body-centered cubic

As a child, were you told the story of the king who asked his three daughters to tell him how much they loved him? The two eldest daughters waxed lyrical, saying that they loved their father as much as gold and the sugar and sweetmeats on their plate. The third daughter said simply that she loved her father as much as she loved the salt on her food. This was not well received by the egotistical king, and the youngest daughter was banished from the kingdom. To cut a long story short, the king tried eating only sweet treats and no savory food for a few days, of which he soon sickened. Realizing the error of his ways—and the true value of his youngest daughter's affection—he welcomed her back with open arms.

This tale is testament to the beauty of simple things. Elements such as gold may steal the limelight visually, but a humble compound such as sodium chloride (table salt) is more important in our daily life. It is believed that, in Roman times, soldiers were paid their salary in salt and that the word "salary" is derived from the Latin word for "salt," *sal*. This is also thought to be the derivation of the term, "worth his salt."

Salt is one of our five main tastes (the other four being sweet, sour, savory, and bitter) and this is why we crave it so much. There are good reasons why we should want salt in our diet. Sodium, like potassium, is an electrolyte and therefore essential for the transmission of nerve impulses and osmotic balance (the regulation of fluid levels in the body). Excess salt, however, creates high blood pressure, which means that the heart has to work harder to pump blood through the body. The recommended intake of salt for an adult is no more than 0.052 ounces (1,500 milligrams) per day. The average salt consumption per person per day in the USA is almost double that.

Pure sodium is a soft, silvery white metal, which is highly reactive and which oxidizes rapidly when cut. It is not found freely in nature and must be prepared from sodium compounds, which—over billions of years— have leached into the seas and oceans from rocks, giving seawater its salty taste.

The chemical symbol for sodium (Na) comes from the Latin word *natrium*, meaning soda. In ancient times, sodium carbonate (also known as soda) was harvested in Egypt from the Natron Valley, where flood water from the Nile would dry out to leave soda crystals, which were used as a cleaning agent; soda even gets a mention in the Bible for this use.

Sodium compounds are very important commercially, in particular in the production of soap, glass, paper, and textiles. Sodium hydroxide is an important alkali in the chemical industry and is used domestically to unblock drains. It has less mundane uses, too: liquid sodium can be used as a coolant in the core of nuclear reactors.

The soft alkali metal sodium is a key ingredient in table salt and cleaning products. In its molten state it is used as a cooling agent inside nuclear power plants.

Magnesium

Category: alkaline earth metal
Atomic number: 12

Atomic weight: 24.3050
Color: silver–white
Phase: solid

Melting point: 1,202 °F (650 °C)
Boiling point: 1,994 °F (1,090 °C)
Crystal structure: hexagonal

Magnesium is essential for almost all living organisms and plays an important role in photosynthesis. It is the central ion of chlorophyll, the green pigment that allows plants to use light to convert carbon dioxide and water into glucose/energy and oxygen. Plants absorb magnesium from the soil and, because the element is a key part of chlorophyll, its deficiency is easy to spot. Leaves start to turn yellow–brown, and red or purple patches can break through as the green pigment fades. A simple solution is to use a magnesium leaf spray, or to add calcium-magnesium-carbonate to the soil.

Humans obtain magnesium from plants, or the animals that ingest them, and a good supply is essential for a number of key processes in the body. Magnesium is the fourth most abundant trace element in our bodies and around 60 percent of it is used to maintain bone structure. The remainder is required for more than 300 biochemical reactions, including those supporting nerve, muscle, and heart function, the regulation of blood sugar levels, the release of energy from food, and protein synthesis.

Magnesium deficiency can develop in sufferers of gastrointestinal disorders such as Crohn's Disease, which affects the absorption of the mineral. Deficiency can also arise if diabetes is poorly controlled and in people who take diuretics regularly. Signs of this deficiency include lethargy, depression, personality change, abnormal heart rhythm, and seizures.

Medical experts are interested in the role that magnesium can play in myalgic encephalomyelitis (ME), which appears to improve when injections of magnesium-saline are administered. Humans need to take in at least 0.007 ounces (200 mg) of magnesium a day and good sources include bran, chocolate, soybeans, parsnips, almonds, and Brazil nuts.

Since the 17th century, the compound magnesium sulfate in the form of Epsom salts has given relief from constipation. The remedy was discovered in 1618 during a drought, when farmer Henry Wicker walked across Epsom Common in southern England and was intrigued to see that cattle would not drink from a pool of water. He tasted the water and found it bitter. The reason for this emerged when he evaporated some of the water and obtained crystals of magnesium sulfate. These were labeled Epsom salts and soon became valued for their medical use, especially in the treatment of constipation. Suspensions of magnesium hydroxide—often called Milk of Magnesia—also became popular as laxatives and as a treatment for indigestion (being antacid, they help to neutralize stomach acid).

Magnesium is also playing a part in reducing the environmental damage caused by cars and airplanes. The element is highly flammable when it is in the form of a ribbon or powder, but as a more solid mass it is very difficult to ignite. This, combined with its strength and lightness when alloyed with aluminum, has led to its use in aircraft and car manufacture. Reducing the weight of cars and airplanes means that they use less fuel, thus reducing pollution—plus the magnesium used can be recycled when the car or aircraft is no longer fit for use.

Spectacular displays of burning magnesium (which flares up with ferociously bright white flame) were a staple of chemistry classes for many of us. It also plays an essential role in the chemistry of the human body, leading to a number of applications in medicine.

Aluminum

Category: post-transition metal
Atomic number: 13

Atomic weight: 26.981
Color: silvery gray
Phase: solid

Melting point: 1,220 °F (660 °C)
Boiling point: 4,566 °F (2,513 °C)
Crystal structure: face-centered cubic

Aluminum is the most abundant metal in the Earth's crust. It is also one of the most useful, being lightweight, durable, easy to work, and simple to recycle. Aluminum's great advantage over other metals is its remarkable ability to resist corrosion. In its pure form it is so reactive that, when exposed to air, it will form a tough, transparent oxide. However, unlike iron, which will rust and gradually wear away, aluminum oxide (alumina) provides a protective coating and is particularly hard. This accounts for aluminum's extensive use in the aerospace, railroad, shipping, and construction industries, and in food containers and cooking utensils.

The name aluminum is derived from the Latin word *alumen*, meaning "alum," which is a naturally occurring salt that was used in Roman times to stop bleeding. It is still in use today in the form of styptic pencils, which help to heal shaving cuts. For many years, aluminum was costly to produce, so the metal that is deemed so commonplace today was considered the preserve of the rich and privileged. At the court of Emperor Napoleon III of France, during the 1860s, visiting heads of state would be served delicacies on aluminum plates to denote their superiority, while mere dukes had to settle for plates of gold.

In its pure state, aluminum is a soft and malleable metal, but due to its highly reactive nature, it is rarely found in this elemental state. Most aluminum is produced from the ore bauxite, and large deposits are found in Australia, Brazil, China, India, and Russia. The process of converting bauxite into aluminum is complex: first aluminum oxide is extracted from the ore, then the aluminum is obtained from the alumina by electrolytic reduction. This is known as the Bayer Process and it requires considerable amounts of electrical energy to power the reactions that take place. In fact, a typical aluminum plant producing around 110,250 tons (100,000 tonnes) a year uses enough electricity to power a small town; every time you recycle an aluminum can, you are saving up to 95 percent of the energy that would be required to make it from ore.

Aluminum is often alloyed with other metals such as copper, zinc, and magnesium to improve its strength. Aluminum is found all around us in the average house: in door handles, window frames, power cables, cooking foil, and compact disks. Its combination of strength and low weight make it ideal for aircraft parts, automobiles, and boats. It is also a good conductor of electricity, and also much lighter than copper, which makes it a good choice for overhead cables.

Large passenger aircraft are hit by bolts of lightning on average once a year—a fact that will strike fear into the hearts of nervous passengers. However, aluminum's conductivity is a bonus in aircraft design. Engineers ensure that electricity can flow freely through the body of the aircraft, so if a bolt of lightning hits the airplane, the current will pass along the outer shell of the fuselage and exit through the tail or wing tips without ever reaching the cabin.

Al ¹³

The stuff of aircraft, food packaging, and garden greenhouses, aluminum is a strong yet lightweight construction material, which conducts electricity and does not rust.

Silicon

Category: metalloid
Atomic number: 14

Atomic weight: 28.0855
Color: metallic and bluish
Phase: solid

Melting point: 2,577 °F (1,414 °C)
Boiling point: 5,909 °F (3,265 °C)
Crystal structure: diamond cubic

Silicon is the second most abundant chemical element in our planet's crust after oxygen. It's the dominant element in flint, which was the first material put to use by humans to make tools and weapons. Indeed, the name derives from the Latin *silicis*, meaning "flint."

Of course, today we've learned to do much more than simply stab each other with pieces of silicon; for proof, take a look out of the window ... or even just take a look *at* the window, where you'll see a big sheet of a silicon-based substance called glass. Silicon is also a key component in semiconductor electronics, household sealants, industrial lubricants, breast implants, and gemstones. If that's not enough, you can thank silicon next time you're on the beach—sand is better known to chemists as silicon dioxide.

Just in case you're not impressed yet, silicon is also the eighth most abundant element in the entire universe and some astrobiologists have suggested that it could form the basis for life on other worlds— much as carbon does on Earth. An atom of silicon can bond with up to four other atoms, just like carbon. However, whereas life on Earth uses water as a solvent, silicon-based biochemistry would require liquid ammonia and extremely low temperatures in order to thrive.

It's hard to imagine a grain of sand or a quartz watch on your wrist being derived from an element that is integral to the physics of dying stars, but this is the cosmic story of silicon. The element is formed by the nuclear fusion process that takes place inside a star. After all the hydrogen has been converted into helium, the star begins a series of nuclear reactions, burning carbon, neon, and oxygen and creating additional elements at each stage; silicon is produced during the oxygen-burning.

Silicon itself can burn via nuclear fusion, during a phase that lasts just two days near the end of the star's life; this activates a process that creates elements such as iron, nickel, and zinc. Iron will not fuse further to make heavier elements, and at this point the star has no more fuel left to power its reactions. It then collapses spectacularly, resulting in some cases in a massive event known as a supernova. These explosions are so powerful and luminous that they can light up an entire galaxy for months, as vast amounts of energy are generated and elements are catapulted into space. This very process, taking place inside stars that formed and died before the Sun, created the silicates from which the planets of our Solar System—and, indeed, the crust of our Earth—are formed.

Silicon was discovered by Swedish chemist Jöns Jakob Berzelius, in 1824. He isolated the first pure sample by heating potassium fluorosilicate with potassium metal. The name, however, was coined by Scottish chemist Thomas Thomson: Berzelius had put forward the name "silicium"; Thomson suggested "silicon" instead after seeing that the properties of the new element were more closely allied to its nonmetal neighbors, boron and carbon. Berzelius had synthesized amorphous silicon—that is, silicon without any kind of crystal structure. However, in 1854,

The semiconductor electronics industry would
be nowhere without the wonder element
silicon, which is essential in the manufacture of
microchips.

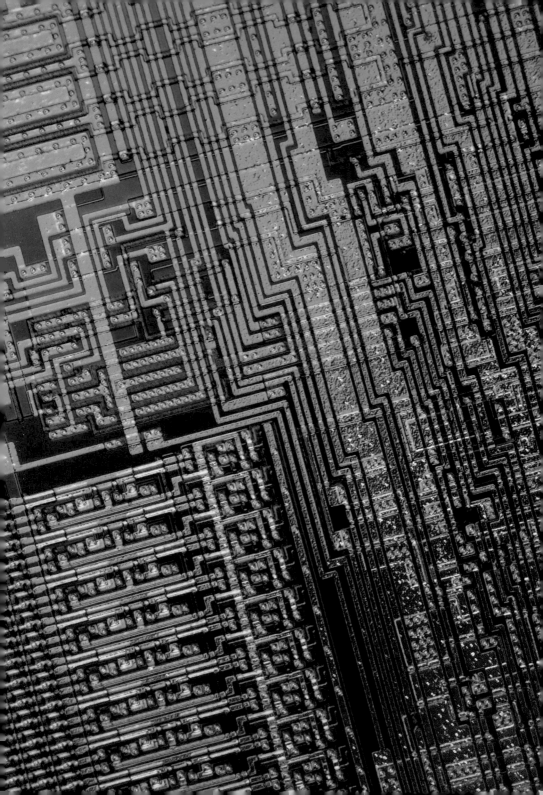

French chemist Henri Deville succeeded in producing crystalline silicon, in which the atoms are arranged into regimented diamond shapes.

Probably the most important use that silicon has ever been put to is in the development of the microchip. Years ago, engineers used to use clunky devices called valves to regulate the flow of electricity. Then, in the early 19th century, they discovered "semiconductors"—materials that are neither perfect insulators nor perfect conductors. One of the first examples was silicon. A sample of silicon can be "doped" with small quantities of other elements to give it an excess of either negatively charged electrons or positively charged "holes"—known as n-type and p-type semiconductors, respectively. Applying a current to a junction of n- and p-type semiconductors creates a transistor (a switch to control the flow of electricity across the junction) and this forms the basis of computers and all other microelectronics. The latest microchips pack billions of transistors onto a square of silicon just half an inch across.

Silicon is mixed with metals, such as aluminum, to improve their ability to form a quality casting (known as "castability"). Silicon mixed with carbon forms silicon carbide, a common abrasive, while quartz crystals are a type of pure silicon dioxide (usually manufactured rather than mined), which resonates at a specific frequency, giving it applications in clocks and watches. Mixed with oxygen, silicon forms the rubber-like polymer silicone, which is used for waterproofing bathrooms and in cosmetic breast implants.

Silicone breast implants were introduced during the 1960s and became very popular with millions of women—and men who had changed sex. During the 1990s, some women claimed that silicone that had leaked from their implants had damaged their health and even caused cancer. This led to court cases and the bankruptcy of a leading US silicone producer. However, subsequent studies disputed the risks of silicone in the human body and this form of implant has continued to be popular.

In December 2011, silicone implants hit the headlines again when a French company, Poly Implant Prothèse (PIP), was found to have been using nonapproved industrial-strength silicone, which is suitable only for use in agriculture, electrical applications, and in mattresses—not the human body. This caused widespread concern because PIP exported thousands of its products abroad, as well as supplying the French market. The industrial-standard silicone implants are not believed to be toxic, although they are more likely to rupture and leak than approved implants. PIP was closed down and the French government recommended that anyone with PIP implants should have them removed.

Biologically, silicon is only required in very small quantities by the human body, where it is used in the skeletal system. Overexposure to certain silicon compounds is known to be hazardous to health—for example, prolonged exposure to magnesium silicate (asbestos) is carcinogenic. However, silicone can be beneficial in the treatment of burns, as it can vastly reduce scarring. Silicone gel sheets can be applied to a scar (not an open wound) to help treat hypertrophic scars (where the scar becomes red, raised, and itchy) and keloids (where the scar continues to grow and encroach over normal tissue). With some silicone treatments, the healing process can be reduced from years to a matter of weeks.

Pictured is a light micrograph of a silicon microchip. Microchips are made from wafers of silicon, "etched" with conducting channels and regions of enhanced positive and negative charge to give them the required electrical properties. Microchips today can house billions of electronic components on an area of just a few square inches.

Phosphorus

Category: nonmetal
Atomic number: 15

Atomic weight: 30.973762
Color: white, black, red, or violet (type dependant)
Phase: solid
Melting point: white: 111 °F (44 °C); black: 1,130 °F (610 °C)

Sublimation point: red: 781–1,094 °F (416–590 °C); violet: 1,148 °F (620 °C)
Boiling point: white: 538 °F (281 °C)
Crystal structure: white: body-centered cubic or triclinic; violet: monoclinic; black: orthorhombic; red: amorphous

---

Phosphorus has been called "the devil's element"—not just because it was the 13th element to be discovered. It was first isolated in 1669 in Hamburg, Germany—the city that, nearly 300 years later during the Second World War, was subjected to one of the most devastating firebombing raids in history when Allied bombers dropped tons of flaming phosphorus, burning the city to the ground. Phosphorus is also highly toxic when ingested, reacting with the liver to cause death within days. Plus it's a central component in nerve gas, which kills by blocking the enzyme that governs the way the body sends signals to vital organs such as the heart. The devil's element, indeed.

Phosphorus was the first element to be found since antiquity. The discovery is a tale in itself. German alchemist Hennig Brand was trying to obtain a philosopher's stone—a mythical substance said to convert base metals into gold. For reasons best known to him, Brand suspected that it might be possible to extract it from urine. To this end, he set about boiling up hundreds of liters of urine—though not until he'd left the stuff for days on end, so it smelt truly dreadful. He believed this would somehow boost his chances of obtaining a philosopher's stone.

Brand distilled the urine down to a white substance, which he found glowed in the dark. Believing that he had actually made a philosopher's stone, he kept the discovery to himself, only selling his secret to others years later. Ultimately, rationality prevailed when, in 1777, French chemist Antoine Lavoisier recognized phosphorus as a new element (the substance produced by Brand had, in fact, been a compound: ammonium sodium hydrogen phosphate). Brand's secrecy cost him the chance to name the element. It was eventually given the name from the Greek *phosphorus*, meaning "light bearer."

Phosphates are the most common form of phosphorus on Earth and consist of phosphorus bonded to atoms of oxygen. Phosphate molecules are of paramount importance in biochemistry, where they are part of ATP (adenosine triphosphate), which acts as a storage medium for energy in body cells. Phosphate is also a component in DNA, and is essential to the skeletal system; bone is made of calcium phosphate. Around 1.5 pounds (0.7 kg) of the average adult human is made of phosphorus. Phosphates, meanwhile, are also essential for the growth of plants and crops, and thus form a key ingredient in fertilizers.

Phosphorus occurs in four natural allotropes, the most common of which are white and red phosphorus. There's also violet phosphorus, formed by heating red phosphorus, and black phosphorus, which is created under high pressure from the other allotropes. White phosphorus is the nastiest form; it catches fire spontaneously in air, leading to its use in tracer bullets and incendiary bombs. If swallowed the substance still burns, causing an unfortunate condition known as "smoking stool syndrome."

P 15

All living things require phosphates, chemical compounds involving the chemical element phosphorus combined with oxygen, in order to function. Pictured is a sample of violet phosphorus.

Sulfur

Category: nonmetal
Atomic number: 16

Atomic weight: 32.066
Color: bright yellow
Phase: solid

Melting point: 239 °F (115 °C)
Boiling point: 833 °F (445 °C)
Crystal structure: orthorhombic

Stink bombs, smelly feet, skunk odor, and bad breath—many of the world's most offensive smells are caused by sulfur. There is even a plant, the titan arum lily, that is nicknamed "the corpse plant" because of its foul, sulfurous odour. Sulfur is smelly as a powder, in crystalline form and, in particular, when it is burned, which is perhaps why it became associated with hell in ancient times.

Brimstone was an early name for sulfur and it was mentioned a number of times in the Bible (most notably when God rained down fire and brimstone to destroy the cities of Sodom and Gomorrah). Particularly zealous priests were said to preach "fire and brimstone" sermons to instill fear of eternal damnation in the hearts of unrepentant sinners.

In reality, sulfur has proved deadly in a number of horrific ways. In around 950 AD, the Chinese made a discovery that would change the world for ever: how to make gunpowder. Sulfur was a key component. Knowledge of this new form of weaponry spread across Asia to Europe, where alchemists refined recipes for it.

Sulfur is also believed by historians to have been a component of "Greek fire," a naval weapon used over centuries by the Byzantine Empire to great effect. The deadly liquid was set alight to create a massive, unextinguishable wave of flame that would advance quickly toward enemy vessels.

Sulfur can exist in different forms: powder, crystal, and as a rubbery solid; the yellow orthorhombic crystalline form is the most common. Its chemistry is complex, due to the different oxidation states in which it can exist. Today, sulfur is an economically important element because it is the raw material of sulfuric acid. This is used in many aspects of industry, including fertilizer manufacturing, oil refining, wastewater processing, removal of rust from iron and steel, and in the production of lead–acid batteries for cars.

Sulfur can be found near hot springs and volcanoes, which is why Sicily was an important supplier for many years. Large deposits of sulfur have been found in the USA, Indonesia, and Japan, although most sulfur is produced by the removal of hydrogen sulfide from natural gas. The gas is run through a tower, which contains a solution of a compound such as ethanolamine; the solution absorbs sulfur compounds from the gas as it passes through. The gas is then free of sulfur and ready for use, while the sulfur by-product is sold on for use in fertilizer.

Hydrogen sulfide is the toxic gas that is said to smell of "rotten eggs." It can be tolerated at low levels, but personal safety gas detectors used by workers in the sewage and petrochemical industries are set to raise the alarm (by changing color) at between 10–15 parts per million (ppm). At levels of 300 ppm or above, it can be deadly. In 2008, a wave of hydrogen sulfide suicides swept across Japan, and in three months 220 people had killed themselves after creating the deadly gas from a concoction of household chemicals. In some cases the lethal toxic gas led to whole apartment blocks being evacuated.

S	16

Sulfur crystals on rock. It's hard to believe that this very same rock is responsible for making flatulence smell—but the compound hydrogen sulfide is the primary culprit for malodorous pant gas.

Chlorine

Category: halogen
Atomic number: 17

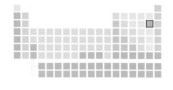

Atomic weight: 35.453
Color: green–yellow
Phase: gas

Melting point: -151 °F (-102 °C)
Boiling point: -29 °F (-34 °C)
Crystal structure: n/a

On April 22, 1915, the German army released chlorine gas at Flanders on the Western Front. The gas carried on the breeze across to the Allied trenches, where it resulted in the agonizing death of 5,000 men and disabilities among 15,000 more. It was later replaced by other chemical weapons such as mustard gas—a carcinogen that penetrates the skin and causes chemical burns. Despite the deadly nature of chlorine, it's intriguing to note that in the last 100 years it has saved many more lives than it has claimed.

Sodium chloride, the most common compound of chlorine, has been used since ancient times, but the gaseous element chlorine was not produced until 1774 by Swedish chemist Carl Wilhelm Scheele (see also oxygen, page 26). Scheele heated hydrochloric acid with a mineral powder containing manganese dioxide, which produced a greenish-yellow gas with a choking smell. Scheele observed a number of key features: that the gas dissolved in water to give an acid solution; it bleached litmus paper; and it attacked most metals exposed to it.

In 1785, a French inspector of dye works, Claude-Louis Berthollet, developed a particularly effective chlorine-based bleach and disinfectant and this is the solution that is used today in homes and swimming pools, with that familiar smell. Bleach made from chlorine was first used to disinfect drinking water in 1897, during a typhoid outbreak in Maidstone, England, and brought the epidemic under control. This led to widespread chlorination of drinking water across the developed world, which has been in place for several decades now and has helped to eradicate devastating waterborne diseases such as typhoid and cholera.

According to the World Health Organization, water chlorination has played a key role in extending life expectancy in the USA from 45 years in the early 1900s to around 77 years in 2012. Across the world, however, 900 million people do not have access to safe water and two and a half billion lack effective sanitation.

Being one of the most reactive elements, chlorine readily binds with other substances. Disinfectants contain chlorine compounds that can exchange atoms with other compounds, such as enzymes in bacteria and other cells. When the enzymes come into contact with chlorine, hydrogen in the enzyme is replaced by chlorine, which causes it to change shape or fall apart. If the enzymes do not function properly, the cell or bacterium will die.

Other main uses of chlorine include chemicals for industry, the manufacture of polyvinyl chloride (PVC), and in solvents. Most chlorine gas is manufactured through the electrolysis of brine, which is obtained through pumping water down bore holes into salt beds.

During the 1980s, chlorine earned a bad reputation again, due to the damaging effect of chlorofluorocarbons (CFCs) on the ozone layer, and in 1987, the Montreal Protocol banning CFCs was signed by leading industrial nations. In 2011, NASA reported that world ozone levels had remained constant over the previous nine years, halting the decline first noticed in the 1980s.

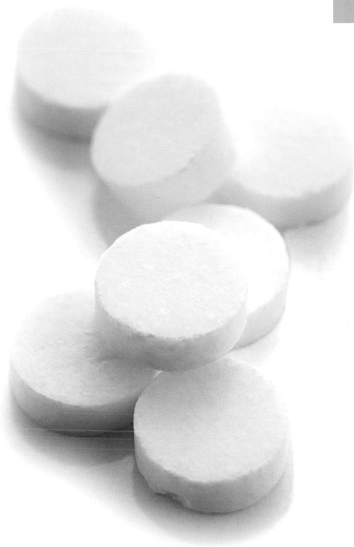

Chlorine tablets are used to purify water for
drinking. These tablets typically contain sodium
dichloroisocyanurate (NaDCC), which when added
to water releases chlorine in low concentrations at
a steady rate. That in turn acts as a disinfectant.

Argon

Category: noble gas
Atomic number: 18

Atomic weight: 39.948
Color: colorless
Phase: gas

Melting point: -308 °F (-189 °C)
Boiling point: -303 °F (-186 °C)
Crystal structure: n/a

Argon is named after the Greek word *argos*, which means "inactive" or "idle." Colorless and odorless, argon is the third most abundant gas on Earth, forming 0.93 percent of the atmosphere. It is a member of the inert or "noble" group of gases, which are thus named because they tend not to bond or associate with other elements. They have a full outer shell of electrons, which gives them little tendency to react chemically. Ironically, it is this very inertness that makes argon suitable for so many important functions in industry.

The gas was officially discovered by the British scientists John Strutt (Lord Rayleigh) and Sir William Ramsay (see also neon, page 32) in 1894. They reported their discovery, but then suddenly fell silent about the new element. This was because the pair had heard of an American competition, in which entrants had to discover something new about atmospheric air. One of the conditions was that the discovery should not have been disclosed before the end of the year. Strutt and Ramsay won first prize, sharing $10,000—the equivalent of around $150,000 today.

Argon had actually been isolated more than a century earlier by Henry Cavendish (see also nitrogen, page 24), during his studies on the chemistry of the atmosphere. Cavendish was perplexed because there was one percent of air that would not combine chemically. Without realizing it, he had stumbled on a previously undiscovered element.

Today, argon is an important industrial gas and is extracted from liquid air.

Atmospherically, we have a plentiful supply, with an estimated 73 trillion tons (66 trillion tonnes) circulating around us. Argon is known for producing the electric blue of illuminated signs, and sometimes a little mercury is added to intensify the color. Conventional incandescent lightbulbs are filled with argon to create an inert atmosphere that prevents the filament from oxidizing at high temperatures.

Argon is important in the steel industry, where it is blown through molten metal, with oxygen (page 26), in the decarburization process. The oxygen reacts with the carbon, forming carbon dioxide, and the addition of argon minimizes unwanted oxidation of precious elements in the steel, such as chromium. Argon is also used as a "blanket gas" in the production of titanium or in welding aluminum, when air has to be excluded to prevent oxidation of the hot metal.

Argon is heavier than air and conducts less heat, which is why it's so useful in double glazing where it forms an effective layer of insulation between the two panes of glass. Its abundance in the atmosphere also makes it a cheap option. Blue argon lasers are used in surgery to weld arteries, destroy cancerous growths, and correct eye defects.

Argon gas in a borosilicate vial. Like the other noble gases, argon is used in discharge lamps, where it is made to glow by passing an electric current through it. Argon lamps give off a purple-blue glow.

Potassium

Category: alkali metal
Atomic number: 19

Atomic weight: 39.0983
Color: silver–gray
Phase: solid

Melting point: 146 °F (63 °C)
Boiling point: 1,398 °F (759 °C)
Crystal structure: body-centered cubic

Potassium is an alkali metal, which makes for great entertainment when it is thrown into water during school chemistry lessons (although don't try this at home). Who could forget the beautiful lilac flames that it produces the second it skims the surface of the water? Putting aside lab pyrotechnics, potassium is one of the most useful and important of elements, without which mankind would die. Given in the wrong doses, however, it can also be deadly.

The name potassium is derived from the English term *potash*, which refers to the method by which it was derived—leaching wood ash and evaporating the solution in a pot to leave a white residue. The chemical symbol K comes from the word *kalium*, which is medieval Latin for "potash." In turn, this may have come from the Arabic word *qali*, meaning "alkali."

Potash salts have been used since ancient times to flavor food, preserve meat, and improve soil. In 1840, German chemist Justus von Liebig made the important discovery that potassium was a necessary element for plant growth and that many soils were deficient. Demand for potash grew rapidly and the subsequent development of fertilizers—using predominately nitrogen, potassium, and phosphorus—has had a huge impact on global population growth.

Agricultural fertilizers use around 95% of the world's potassium production, most of which is in the form of potassium chloride. Wood ash was initially used as a source for fertilizer before natural deposits of potassium chloride were found, notably in Germany and Canada.

Potassium is vital for human life and health. It is an electrolyte (or ion) and carries electrical impulses across cells. It's essential in nerve transmission, the regulation of the osmotic balance between cells and their surrounding fluid, and kidney function.

If you exercise heavily, you lose electrolytes in your sweat, particularly sodium and potassium. Also, a severe bout of diarrhea can decrease potassium levels. Cells with high electrical activity, such as muscles and nerves, are particularly affected by potassium deficiency and this can lead to weakness, cramps, and confusion. The electrolytes must be replaced to keep the concentrations of body fluids constant, which is why sports drinks and oral rehydration treatments contain sodium chloride (table salt) and potassium chloride. Rich dietary sources of potassium include bananas, apricots, lentils, broccoli, cantaloupe melon, and peanuts.

Excessive levels of potassium in the body can have severe effects on the central nervous system and an excess of potassium chloride may result in diarrhea and convulsions. If a concentrated solution of potassium chloride is injected into the body, this affects the chemical balance of body cells, most notably those of the heart, which stops beating. Potassium chloride has been used by medical teams as a means of euthanasia for terminally ill patients and in capital punishment as a lethal injection, where it's usually administered with a barbiturate (to render the prisoner unconscious) and a muscle relaxant (to paralyse the diaphragm and respiratory muscles).

A sample of the soft, alkali metal potassium. The element plays a major role in the human body, mediating the transfer of signals through the nervous system. Meanwhile, the compound potassium nitrate is the main ingredient of gunpowder.

Calcium

Category: alkaline earth metal
Atomic number: 20

Atomic weight: 40.078
Color: silver–gray
Phase: solid

Melting point: 1,548 °F (842 °C)
Boiling point: 2,703 °F (1,484 °C)
Crystal structure: face-centered cubic

It may seem odd to consider calcium in its pure form as a silvery gray metal, when many of us associate it with white, chalky things, such as bones, teeth, and England's white cliffs of Dover. This is due to the fact that, like many elements, calcium is rarely seen in its pure form. This is because calcium is unstable and, when exposed to air, quickly forms calcium hydroxide and calcium carbonate (which forms those famous white cliffs). When placed in water, calcium metal reacts to produce hydrogen gas, although not with the fireworks or drama of the alkali metals.

Calcium is essential to almost all living things, and forms the "building blocks" of the human skeleton (as calcium phosphate) and the shells of mollusks and crustaceans (as calcium carbonate). Bone is not static, as many people believe; it changes, is broken down, and rebuilt constantly, and most of the calcium that we absorb is required to support this process. Children, pregnant women, and elderly people all require between 1,000 milligrams (0.035 oz) and 1,500 milligrams (0.053 oz) of calcium each day, and healthy adults are advised to take in at least 0.035 ounces (1,000 mg). Rich sources include dairy products, milk chocolate, vegetables such as broccoli and cabbage, and red kidney beans.

Calcium is also required for other processes in the body, and if a calcium deficiency arises, the body will take what it needs from bone. This must be replaced and as we get older the process does not completely cover the loss, which can lead to problems like osteoporosis. For calcium to be absorbed, we must have a sufficient supply of vitamin D, which is found in fish oil, seafood, eggs, and some dairy products.

Calcium oxide, or lime as it's more familiarly known, has been used since early times to make building mortar, and this is well documented in texts from Roman times. Mortar was created by mixing lime with sand and water to produce a cement. Over time, this cement would harden as it absorbed carbon dioxide from the air to form calcium carbonate. Calcium is named after the Latin word *calx*, which means "lime."

Limestone is prized for the warm, honey-colored glow it casts and one of the finest architectural examples is the UNESCO World Heritage city of Bath, England. This Georgian city, renowned for its uniform beauty and elegance, is constructed from locally quarried limestone, which formed during the Jurassic period more than 144 million years ago.

Lime-rich soils are also known as chalky or alkaline soils, and present a challenge to the gardener because of their dryness and lack of nutrients. However, if soil is too acidic, this can prevent plants from absorbing the nutrients they need—so lime can be added to raise the pH level of the humus.

There are many calcium ores, the main ones being calcite, anhydrate, dolomite, and gypsum. World lime production far exceeds production of calcium metal. As well as being an ingredient in cement, lime is also used as a water treatment, a fertilizer, and in steelmaking as a flux, where it bonds with impurities in the ore to create a slag that can then be removed.

Calcium is the fifth most abundant element in the Earth's crust. Calcium metal, the pure form of the element, is quite different to the calcium compounds, such as bone and limestone, with which we are more familiar.

Scandium

Category: transition metal
Atomic number: 21

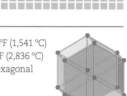

Atomic weight: 44.955912
Color: silver–white
Phase: solid

Melting point: 2,806 °F (1,541 °C)
Boiling point: 5,136 °F (2,836 °C)
Crystal structure: hexagonal

The name gives away at least some of the history behind this element, which can be found at the top of the periodic table's group 3. Scandium was first discovered during analysis of a sample of minerals gathered in Scandinavia.

Its existence didn't come as too much of a surprise—at least not to everyone. Dmitri Mendeleev, the Russian chemist who first formulated the periodic table in 1869 (see pages 8, 232), speculated that there should be another element to fill the gap in the table between calcium (atomic number 20) and titanium (atomic number 22). He referred to this elusive substance as "eka-boron." The name derived from *eka*, the Sanskrit prefix meaning "one," and "boron" (an element that Mendeleev had originally classified at the top of group 3 and thus one place away from where eka-boron should lie). It was just ten years later that Lars Fredrik Nilson, of the University of Uppsala in Sweden, first isolated scandium from a sample of the complex mineral euxenite.

Even then, only the tiniest trace amounts of a white solid called scandia, or scandium oxide, were made. Nilson used spectrometry to confirm that he had found a new element. This is a technique where a substance is burned and the light given off is passed through a prism to split it into its constituent colors. Some of these colors appear as darkened bands—where the atoms of the element have absorbed specific wavelengths of light—and the pattern of bands is unique to each element. Sure enough, Nilson's sample absorbed light at a range of colors never before seen.

It would be much longer before a solid lump of scandium metal was finally produced, in 1937. Part of the problem was that scandium is spread so thinly throughout the Earth's crust. In contrast with other mineable elements, such as iron and gold, there are relatively few concentrated deposits of scandium ore. Instead, a large volume of material needed to be mined and processed in order to produce just a small quantity of this silver–white metal. The situation is much the same today and, for this reason, global production of scandium totals just a few tens of kilograms per year.

Scandium has nonetheless found commercial use. Adding just a tiny quantity (as little as 0.1 percent) to aluminum, for example, greatly improves the strength and other mechanical properties of the resulting alloy—so much so that scandium–aluminum alloys have been used in aircraft manufacture, in particular the Russian Mig-29 fighter. They have also found applications in high-performance sports equipment, such as baseball bats and bicycle frames. Meanwhile, solutions of pure scandium speed the germination of seed crops.

Another surprising application of scandium is in discharge lamps. These work by charging up vapor with electricity, which is subsequently emitted as light. Adding scandium to the vapor softens the quality of the light, making it more like natural daylight and thus a good choice for studios and sporting venues. Although scandium is rare on Earth, it is plentiful in outer space—having been detected in the light from the Sun, as well as more distant stars.

Sc 21

A sample of scandium. The transition metal can be added to aluminum to make an exceptionally strong alloy, used in aircraft construction.

Titanium

Category: transition metal
Atomic number: 22

Atomic weight: 47.867
Color: silver
Phase: solid

Melting point: 3,034 °F (1,668 °C)
Boiling point: 5,949 °F (3,287 °C)
Crystal structure: hexagonal

Titanium is the ninth most abundant element on Earth—which is lucky, given the wealth of applications we seem to have found for it. As strong as steel but 45 percent lighter, and all the while unaffected by the deadly phenomena of metal fatigue and cracking, it's no surprise that this metal from group 4 of the periodic table has found countless applications in aviation. Aircraft frames and especially jet engine components all use titanium. A single Boeing 777 airliner is estimated to have 65 tons (59 tonnes) of it.

Titanium is not only strong and light, but also resistant to corrosion; it corrodes ever so slightly, but rather than flaking off like rust on iron—exposing the metal beneath to yet more corrosion—the oxide effectively encases the metal in a protective film that prevents further corrosion from taking place. The thickness of the film is initially just a couple of nanometers, but grows to a maximum of around 25 nanometers after four years. This ability to thwart the highly corrosive effects of seawater has led to many maritime uses for titanium, such as propeller shafts, oil rig supports, and the outer hulls of submarines.

Titanium's reluctance to corrode means that it is highly inert and nontoxic. This, coupled with its strength, makes it an ideal choice for prosthetic body parts. In particular, it's used in replacement hip joints, pins for screwing broken bones back together, and plates for mending fractured skulls. These components last in the body for decades. Titanium is also special in this regard, because it seems to integrate well with living bone—believed to be due to an affinity between bone and the thin oxide film that coats the metal. For this reason, all medical titanium is first treated with high-voltage electricity, which removes all surface coverings and allows a fresh protective layer of oxide to form.

Sports goods manufacturers are fond of using titanium in equipment such as golf clubs, tennis racquets, and bicycles. Lightweight helmets and even horseshoes have been crafted from it. However, it's not just the pure metal that's useful: its compounds have found a variety of applications. For example, titanium dioxide is extremely white and opaque and is therefore used as a pigment and as an opacity agent in paint, in paper (to stop ink from the other side from showing through), and in sun creams (to stop ultraviolet rays). Recently, titanium has also found an unexpected application as a self-cleaning coating for windows. Meanwhile, titanium and nickel make up the shape-memory alloy nitinol—which allows you to accidentally sit on your spectacles, only for them to spring back into shape once you get up.

William Gregor, a vicar and amateur geologist from Cornwall, England, discovered titanium in 1791. He found some black sand by a stream from which he was able to extract the oxide of a new element. He named it menachanite after the parish where it was found. In 1795, German scientist Martin Heinrich Klaproth independently discovered the element. Klaproth gave it the name by which it goes today (after the "Titans" of mythology)—perhaps he foresaw what a mighty element titanium would become.

The silver-colored and brittle transition metal
titanium. A strong, lightweight metal, titanium is
used extensively in aerospace engineering. The
compound titanium dioxide is used as a pigment
in most white paints.

Vanadium

Category: transition metal
Atomic number: 23

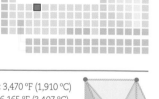

Atomic weight: 50.9415
Color: silvery gray
Phase: solid

Melting point: 3,470 °F (1,910 °C)
Boiling point: 6,165 °F (3,407 °C)
Crystal structure: body-centered cubic

Vanadium is one of the elements that has played a key role in modern history, changing the way we live, work, travel, and influence our environment. In pure form, it is a hard, shiny, silvery metal and the formation of an oxide layer gives it good resistance to corrosion. When alloyed with steel, vanadium makes the alloy much stronger and lighter, which is why it became so useful during the 20th century.

Henry Ford, of Model T car fame, said, "But for vanadium there would be no automobiles." Ford's objective was to create a motor car for the masses, which meant sourcing good, durable materials at low cost. During a car race in Florida, Ford came across vanadium in the wreckage of a French vehicle. Examining part of a valve spindle, Ford was surprised at how hard yet lightweight it was. The valve spindle was made from a steel that contained a few percent vanadium, an alloy that was not made in the USA at the time. This did not deter Ford, who decided this was the perfect material for his inexpensive mass-market motor. Production using vanadium began in 1913 and, prior to this, few Americans could afford to run a motor car. When the Model T was discontinued in 1927, more than 15 million had been made, with many components comprising vanadium steel. Ford played a huge part in a revolution that was to have an immense impact on life in the Western world … and on climate change.

Vanadium also played a key role during the First World War, as it enabled fighter planes to carry powerful but light cannons rather than less effective machine guns

and because vanadium steel offered greater protection against bullets than earlier alloys. It became a standard ingredient in helmets and other forms of protective armor.

Today, most vanadium is used as ferrovanadium, which is added to steel to toughen it. You can find it around the house in drill bits, routers, pliers, and other tools, some of which may be marked with "vanadium."

Vanadium was actually discovered twice. In 1801, Andrés Manuel del Río, a Mexican professor of mineralogy, found vanadium in a brown lead ore and was intrigued by the many colors of its salts. He remarked that it was very similar to chromium, and a French chemist who examined a specimen decided that it was indeed a chromium mineral. Thirty years later, Nils Gabriel Sefström, a Swedish chemist, discovered vanadium in a sample of cast iron from ore mined at Småland. The workers there had thought it odd that their cast iron was at times strong and at other times brittle. Sefström chose the name vanadium, from Vanadis, the Scandinavian goddess of beauty.

Vanadium is an essential element for the human body, although a mere 40 micrograms a day should be sufficient to meet our needs. It is believed to regulate the enzymes that control sodium in the body, and tests on rats and chickens have shown that vanadium promotes growth, which may be the same for humans. Vanadium is abundant in most soils and is absorbed by a variety of plants and fungi; toadstools, in particular, accumulate a large amount of vanadium, as do marine worms.

23
V

Vanadium is used industrially to improve steel and it has good corrosion resistance to alkalis, sulfuric acid, hydrochloric acid, and saltwater.

Chromium

Category: transition metal
Atomic number: 24

Atomic weight: 51.9961
Color: silvery gray
Phase: solid

Melting point: 3,465 °F (1,907 °C)
Boiling point: 4,840 °F (2,671 °C)
Crystal structure: body-centered cubic

Chromium is a hard, lustrous, silvery transition metal that is resistant to oxidation. Its name is derived from the Greek word *chroma*, meaning "color." At one time, the pigment chrome yellow (lead chromate) was a favorite on the artist's palette due to its great intensity, and it was also used to paint American school buses bright yellow, as that made them easier to see in the semidarkness of the early morning. It was replaced by cadmium yellow and orange when it was realized that chrome yellow contained a toxic, heavy metal.

The element was discovered in 1797 by the French chemist Nicolas Louis Vauquelin, who also discovered beryllium. Vauquelin was made a member of Paris's prestigious Academy of Sciences in 1791 at the age of only 28. He left Paris for a time to escape the worst of The Terror of the French Revolution and on his return became an inspector of mines and a professor at the national school of mines. He became fascinated by a bright red mineral found in Siberia in 1766 (now termed crocoite), examined a sample, and made a solution from it. First, he precipitated lead and then he isolated chromium from what was left. The new element could produce a marvelous array of colors in solution, hence Vauquelin's choice of name. He also found that green emeralds are formed by a tiny amount of chromium in corundum, the mineral form of aluminum oxide.

The main ore of chromium is chromite and key sites for mining include South Africa, Zimbabwe, and Finland, with further deposits to be exploited in the USA, Russia, and Greenland. Chromium is the 21st most abundant element in the Earth's crust.

Chromium is essential to human beings as a trace element and is thought to influence how the hormone insulin functions in the body and the release of energy from food. Recommended daily intake for a healthy adult is between 25 and 35 micrograms per day and good dietary sources include meat, oysters, whole grains, lentils, eggs, and kidneys. The organ with the largest amount of chromium in the body is the placenta. Animals also require chromium. If they become deficient in the element, their cholesterol levels decrease and they also develop diabetes.

Chromates can be particularly toxic to humans and exposure to some chromium compounds can lead to a nasty skin complaint. In the early 19th century, workers in Scotland handling chromium compounds began to develop ulcers that exposed raw flesh and caused terrible itching. These became known as chrome ulcers. People working in chrome-plating, dyeing and chrome-tanning industries also became susceptible.

Today, most chromium is alloyed with iron and nickel to harden stainless steel—some of which consists of almost 25 percent chromium. The element is also used as a thin layer in plating to create a shiny, hard surface that offers good resistance to corrosion. Chrome plating became extremely popular for car bumpers during the 1950s and 1960s, adding literally dazzling glamor to Pontiacs, Chevys, Cadillacs, and many more.

Cr 24

Chromium is a hard, metallic element. Many chromium compounds are intensely colored and the ore crocoite, also known as Siberian red lead, is prized for its use as a red pigment in oil paints.

Manganese

Category: transition metal
Atomic number: 25

Atomic weight: 54.938049
Color: silvery gray
Phase: solid

Melting point: 2,275 °F (1,246 °C)
Boiling point: 3,742 °F (2,061 °C)
Crystal structure: body-centered cubic

Did you know that prison bars are made of a type of steel that contains around 13 percent manganese? The reason why this alloy is used in prisons, and also for railroad tracks, helmets, and safes, is because of its extreme strength. The alloy was patented in 1883 by the 24-year-old Briton Robert Hadfield, and is sometimes known as Hadfield steel.

Manganese in its pure form is a hard and brittle metal. Mankind has been using forms of manganese for at least 17,000 years; our ancestors would express their creativity in cave paintings using black manganese oxide pigments (with red iron oxide). Manganese minerals were also used in the glassmaking process and the Roman philosopher and naturalist Pliny the Elder, who died following the eruption of Vesuvius in 79 AD, described a powder used to remove other elements and make glass clear.

The pure element was isolated in 1774 by the Swedish scientist Johan Gottlieb Gahn. The name manganese is derived either from the Latin word *magnes*, which means "magnet" (and the element's common mineral pyrolusite does have slight magnetic attraction), or from magnesia nigra, the name for the black rock found in Magnesia, Greece.

Manganese is essential for all species, but only in trace amounts for human beings. It is thought to help form and activate some enzymes and is involved in the metabolism of glucose. Many people get a plentiful supply of manganese from the infused drink tea and also bread, nuts, cereals, and vegetables. Manganese is stored mostly in our large bones, but also in the liver, pancreas, and the pituitary and mammary glands.

Like many elements that we need in trace amounts, exposure to certain forms or large amounts can be hazardous. Manganese dust or fumes can be toxic and miners who were exposed to such risks could develop "fume fever" or even "Manganese madness," whereby those afflicted would laugh or cry involuntarily, become aggressive, and suffer hallucinations.

The most important manganese ore is pyrolusite and the main mining areas are in South Africa, Russia, Ukraine, and Australia. It is most plentiful on the ocean floor, however, where there is estimated to be a trillion tons of manganese "nodules" scattered over vast areas of the northeast Pacific in particular.

This marine wonder was part of an elaborate Central Intelligence Agency (CIA) hoax in 1974, when the entrepreneur Howard Hughes sparked a "manganese fever" of his own by commissioning a ship to probe the ocean floor northwest of Hawaii and harvest these potentially valuable nodules. This was a cover, however, as Hughes had been hired by the CIA to raise the Russian missile submarine K-129, which sank there in 1968. The mission was a costly failure; during the raising of the sub, the section that held the code books (which were the main goal of the operation) broke off, and Hughes's team were left with a few missiles and the bodies of six Russian crewmen, who were buried at sea with military honors.

25
Mn

The hard, silvery gray transition metal manganese.
Steel that contains manganese is extremely tough
and strong—so strong that it has been used in
mining machinery and soldiers' helmets.

Iron

Category: transition metal
Atomic number: 26

Atomic weight: 55.845
Color: silvery gray
Phase: solid

Melting point: 2,800 °F (1,538 °C)
Boiling point: 5,182 °F (2,861 °C)
Crystal structure: body-centered cubic

It's impossible to avoid superlatives when writing about iron: it's the most abundant element on Earth (if you include the planet's crust and molten core); it generates Earth's magnetic field, protecting us from harmful cosmic radiation; it's the heaviest element that can be made in the nuclear fusion that takes place inside stars; and, without it, industry would collapse … literally.

Iron was first smelted by the Hittites of Asia Minor (modern-day Turkey) around 1500 BC. They kept the secret to themselves but when they were invaded in 1200 BC, the ironmakers fled far and wide, taking their valuable skills with them. So began the Iron Age, when important advances in toolmaking and building represented one of the largest leaps forward in civilization across the world.

For centuries afterward, a warrior carrying an iron sword was invincible. However, the Roman philosopher Pliny showed remarkable insight when he stated, "with iron, wars take place, and not only hand-to-hand but from a distance, with winged weapons launched from engines." Pliny had a vision of the future where iron cannon balls (allied with gunpowder) would become a devastating force. During the two world wars of the 20th century, iron in the form of bombs and shells killed millions of people and almost flattened whole towns and cities. Like fire and water, iron can be a good servant but a poor master.

So, what is it that gives iron its legendary strength? The answer lies in the size of the grains in iron's crystal structure. Iron has a cubic crystal structure, which is inherently strong. However, a sample of iron in its pure form has a granular structure—made of many grains in which the crystal is aligned in different directions. The boundaries between these grains can produce defects that weaken the iron. However, working the iron at high temperature (as is done in foundries) serves to make the grains smaller, leading to a more consistent structure and producing metal that is stronger and tougher.

Iron is formed by the nuclear fusion process that takes place within stars. The Earth's 4,350-mile (7,000-km) diameter core consists mostly of molten iron and the 1,553-mile (2,500-km) wide central core is believed to consist of solid iron. The flow of molten iron within the core creates electric currents that in turn create Earth's magnetic field, which reaches tens of thousands of miles out into space, protecting us from harmful solar radiation and the solar wind that could strip away our atmosphere.

Earth's core is like a giant magnet with two main poles: north and south. Compasses—which were first used in ancient China and became maritime navigational aids in the 12th century— are minimagnets in themselves, with the needle pointing to magnetic north. Before compasses came into use, sailors charted their course using the stars.

It's not just mankind that has used Earth's magnetic field as a navigational aid. Birds and other creatures find their way across continents and oceans by sensing the direction of the magnetic forces

26

Fe

A rare sight of iron in its rust-free state. Iron is a true building block of life—from the blood in our veins to the stainless steel of your flatware, kettle, and car.

around them. Scientists believe that birds can actually see Earth's magnetic field because their eyes contain molecules linked to the part of the brain that processes visual information. This built-in "sat-nav" also enables the birds to gauge the best places to stop and feed on their huge migratory journeys.

Iron may be crucial in so many ways on Earth, but this leviathan of the periodic table does have one weakness when it comes to its usefulness to humans: it reacts with oxygen to form iron oxide, or rust. This can be overcome by coating it with tin (see page 120) or galvanizing it with zinc (see page 76)—which is just as well, because iron's use in steel is so important globally. In 2010, 2.6 billion tons (2.4 billion tonnes) of iron were mined across the world, mostly in China, Australia, Brazil, and India. The main ores are magnetite, goethite, and hematite.

One of the most breathtaking sights in the world has been formed over eons at Bryce Canyon in Utah, USA, where the rocky landscape glows with the pinky red hues of hematite. Curious pinnacles pierce the landscape, known as "hoodoos," "fairy chimneys," and "earth pyramids." They were formed by water erosion and often look like fiery totem poles, reflecting the Sun's glow as it sets.

The Carajás mine in northern Brazil is the world's largest iron ore mine; vast swathes of rainforest had to be dug up to create it. However, after the iron ore has been mined, the owners aim to restore the rainforest vegetation and in order to do so thousands of saplings are being cultivated to recreate the forest canopy. The test will be whether, given the disruption to the soil and topography, the rainforest can be artificially restored to its original glory.

Around 95 percent of iron is produced by heating iron ore with coke (pure carbon) and limestone in a blast furnace, to produce pig iron. This is the raw material of steel, which contains 1.7 percent carbon to make it less brittle and more resistant to corrosion. Nothing can rival steel for strength, durability, and economy.

Iron can be cast, machined, and welded into myriad forms, which makes it the most versatile of metals. It accounts for 90 percent of all the metal that is refined globally, from vast bridges, skyscrapers, and container ships to your family car, screwdriver, and the paper clip on your desk. Iron is all around us, in many shapes and forms.

It is also pumping through our arteries and veins in hemoglobin, a protein that contains iron atoms and transports oxygen to body cells. If we become deficient in iron, the body makes fewer red blood cells and we become anemic, suffering tiredness, lethargy, and breathlessness. Good dietary sources of iron include red meat, liver, dried fruit, bread, and eggs. If you fancy something more indulgent—and in moderation, of course—the Italians have an old saying that may appeal: "red wine makes good blood."

Iron, and steel (which is an alloy of iron and carbon), are both susceptible to corrosion—an "oxidation" reaction, where iron reacts with oxygen to form iron oxide, also known as rust. The steel hulls of ships can be protected from corrosion by being "galvanized" with a zinc coating.

Cobalt

Category: transition metal
Atomic number: 27

Atomic weight: 58.9332
Color: metallic gray
Phase: solid

Melting point: 2,723 °F (1,495 °C)
Boiling point: 5,301 °F (2,927 °C)
Crystal structure: hexagonal

Curiously, this element, which is synonymous with an intense and beautiful shade of blue, is named after *kobald*, the German word for "goblin." It was given its nickname by 16th-century silver miners in Saxony, Germany, who smelted what they thought was a silver ore and were most disappointed to find that—instead of a cache of precious metal—all they produced were toxic fumes of arsenic. What they *had* found was cobalt arsenide, and they cursed it, believing it had been bewitched by goblins.

Cobalt was more highly prized in ancient times, particularly in China, for pottery glazes, and in Egypt, where a glass object colored with cobalt was found in the tomb of King Tutankhamen (*c.* 1352 BC). Before cobalt was formally discovered, the material used for coloring glass or pottery was called "zaffer" or "smalt" and was made by roasting the mineral ore smaltite with sand to form cobalt silicate, which was ground up into a fine powder.

The element itself was discovered by Swedish chemist Georg Brandt in the 1730s. Brandt analyzed dark blue ore from a copper mine in Västmanland and deduced that it contained a previously unrecognized metal. Chemists at the time disputed Brandt's findings, arguing that the "metal" was a compound of iron and arsenic. Eventually, however, Brandt was proved right and he named the element after those disgruntled miners and their noxious compound.

Cobalt is behind a rather pleasing trick of chemistry—"invisible ink," which has been used in espionage since the 17th century. To create the "ink," cobalt ore is dissolved in acids to produce cobalt chloride crystals, which are then dissolved in water with glycerol to form an almost colorless solution. The "invisible" message is written, then another innocuous letter is written on top of it at a different angle and using standard ink. To make the "invisible" letter legible, all the recipient needs to do is heat the paper. This drives off the water and glycerol molecules from the cobalt and allows chloride ions to move in, giving the ink a dark blue color.

In the 20th century, cobalt carried quite a reputation, due to its highly radioactive isotope Co-60, which was a deadly component of the fallout from above-ground nuclear testing in the 1950s. Ordinary cobalt is not radioactive at all, however. In fact, it is a key constituent of vitamin B12 and essential for the normal functioning of the brain and nervous system, and the formation of blood. Deficiency in this nutrient leads to pernicious anemia, where the body is unable to produce sufficient red blood cells to carry all the oxygen that it requires. Foods rich in vitamin B12 include eggs, liver, salmon, herring, and sardines. Vegans cannot obtain it from their diet, so it's essential that they take a supplement.

Cobalt is in low supply in the Earth's crust, and most is mined in the Democratic Republic of Congo, Zambia, Australia, Russia, and Canada. A large percentage of cobalt is used to create superalloys—metals whose stability at all temperatures makes them ideal for use in blades for gas turbines and jet aircraft engines.

Co
27

Cobalt is an important catalyst—an element or
compound that facilitates a chemical reaction
while remaining unchanged itself—used in the
refining of liquid fuels such as petroleum.

Nickel

Category: transition metal
Atomic number: 28

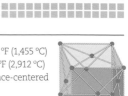

Atomic weight: 58.6934
Color: silvery white
Phase: solid

Melting point: 2,651 °F (1,455 °C)
Boiling point: 5,274 °F (2,912 °C)
Crystal structure: face-centered cubic

Much of the nickel mined on Earth may have arrived via gigantic meteorites, one of which (thought to be around the size of Mount Everest) slammed into Ontario, Canada, 1.8 billion years ago. Meteorites contain iron and nickel, and at least 30 percent of the world's nickel is mined in the area of Sudbury, where the Canadian meteor hit. Not all of the nickel deposits at Sudbury necessarily came from the meteor, however, as much could have welled up from the Earth's interior. Nickel is also mined in Russia, South Africa, Australia, and Cuba.

There is evidence of nickel being used in early times mixed with iron (page 66) and worked into tools and weapons. The name is derived from the German word *kupfernickel*, which means "devil's copper"—more correctly, "St Nicholas's copper." This was the term given to a reddish brown ore found by German copper miners. They could do little with it other than use it to color glass green.

Mineralogists tried in vain to extract copper from the ore, but progress was made in Stockholm in 1751, when Axel Fredrik Cronstedt studied a new mineral found in a cobalt mine at Los in Sweden. Instead of copper, however, he discovered an unknown metal, which he found he could also isolate from kupfernickel. Many chemists believed that Cronstedt had found an alloy, so he had to wait four years before his discovery of the element nickel was recognized.

Nickel in its pure form is a silvery white, lustrous, ductile metal that resists oxidation. At least half of the nickel produced today is alloyed with steel to make corrosion-resistant stainless steel. This has myriad uses— from the automotive and aerospace industries to the flatware and cookware we use at home. Nickel has also been used in coinage, industrial catalysts, and, because it can resist high temperatures, in gas turbines and rocket engines.

Other alloys of nickel have some remarkable uses: Inco alloy 276 consists of 57 percent nickel with 16 percent each of chromium and molybdenum, plus small amounts of other metals. Unlike stainless steel, this alloy can resist corrosion from hydrogen sulfide gas and it is therefore used in places where that gas often occurs, such as deep wells in Earth's crust.

Nickel is essential for some species and the average human daily intake of 150 micrograms is considered to be more than sufficient. If you enjoy a cup of tea you'll certainly meet your daily requirements, with dried tea leaves containing 0.0001 ounce per pound (7.6 mg of nickel per kg).

The element can also be a health hazard, however. Breathing in nickel dust can be dangerous and nickel carbonyl gas is lethal in small quantities. These risks are mostly limited to people who work in nickel mining and industry. The element can, however, cause skin irritation to some people who are particularly sensitive to it. Nickel in stainless-steel watches, jewelry, and spectacle frames can cause itching and dermatitis, which is worsened by the acid in sweat dissolving some of the metal. On a more positive note, products containing nickel can be reused many times over, which makes it one of the world's most recycled metals.

Ni 28

Nickel is a silvery white, hard, and ductile transition metal. It's one of the world's most recycled metals and used in coinage in many countries.

Copper

Category: transition metal
Atomic number: 29

Atomic weight: 63.546
Color: red–orange
Phase: solid

Melting point: 1,985 °F (1,085 °C)
Boiling point: 4,644 °F (2,562 °C)
Crystal structure: face-centered cubic

Copper is one of the most important metal elements in history and was one of the first to be mined. It's estimated that copper has been in use for 10,000 years; a copper pendant has been found in modern-day Iraq that dates back to 8700 BC. Copper's durability, as well as its alluring color, has undoubtedly enhanced its appeal over the ages; when archeologists were excavating the Great Pyramid of Giza in Egypt, they found a portion of a water plumbing system —the copper tubing used in it was still in serviceable condition after 5,000 years.

Copper played a huge part in the development of civilization during the Bronze Age (3000–1000 BC), when it was alloyed with tin to create that era's eponymous metal, which was ideal for tools and weapons. In Roman times, bronze became used for everyday items such as coins, razors, musical instruments, jewelry, and containers, which created huge demand for copper and a thriving trade around the Mediterranean Sea.

Copper's name is derived from the Latin word *cyprium,* after the island of Cyprus, which was a key exporter of the element in early times. The name was later shortened to *cuprum,* which is from where copper derives its chemical symbol.

Alongside gold and silver, copper belongs to group 11 of the periodic table, and is one of the transition metals. These elements share certain attributes: they are soft and malleable, highly ductile (easy to stretch), and are good conductors of electricity. After silver, copper has the highest electrical conductivity of any metal, and its use in vast quantities of copper wiring makes it as important today as it was in ancient times. Like gold and silver, copper was used for coinage (hence the British term "coppers," meaning small change coins), but it was the most common and therefore the least valuable of its group. Most copper today is used in electrical equipment, roofing and plumbing, heat exchangers, and in alloys— notably with zinc to form brass.

Copper weathers well to form verdigris or patina (copper carbonate), which is corrosion-resistant and has been used in paintings, buildings, and statues for many years, including the 18th-century Copper-Roof Palace in Warsaw, Poland, and the Statue of Liberty in New York Harbor, USA. Copper plating has also been used on the hulls of wooden ships to keep them free of barnacles, seaweed, and ship worms and thus improve speed and maximize the hull's longevity. The ships Christopher Columbus sailed in the 15th century were among the first to have this feature.

Copper is essential for the survival of all organisms. In the human body, it combines with proteins to produce enzymes. These act as catalysts for the release of energy from cells, the transformation of melanin for skin pigmentation, and the maintenance of connective tissues (especially in the heart and arteries). An adult requires 0.0004 ounces (1.2 mg) of copper a day and good sources include seafood, meat, almonds, sunflower seeds, and bran.

Unmixed native copper in its oxidized state. The
corroded parts have a green hue. Copper is one
of the most-used metal elements we have—with
applications in plumbing, electrical wiring, coinage,
and many branches of engineering.

Zinc

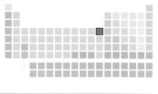

Category: transition metal
Atomic number: 30

Atomic weight: 65.409
Color: bluish white
Phase: solid

Melting point: 788 °F (420 °C)
Boiling point: 1,665 °F (907 °C)
Crystal structure: hexagonal

Zinc is an essential trace element for humans, other animals, and plants. More than 200 types of enzyme contain zinc, and these help to control our development, growth, digestion, immune system, and fertility. Semen is rich in zinc, and scientists believe that low concentrations of this element can result in reduced sperm counts in men.

Good dietary sources of zinc include red meat, liver, cheese, oysters, brewer's yeast, maple syrup, and bran. Zinc deficiency in humans can lead to stunted growth and sexual development, which was highlighted in 1968 by the acclaimed Indian biochemist Ananda Prasad, who studied zinc deficiency in humans and animals. Prasad became the world's foremost researcher into human zinc deficiency, and carried out extensive studies in Egypt, where zinc is scarce in the soil and many people suffered from stunted growth as a consequence. Prasad and his team found that supplementing the diet with zinc sulfate was the most effective cure.

Low levels of zinc in soil is one of the most common micronutrient deficiencies in agricultural land and is particularly problematic for cereal crops. Without adequate quantities of zinc, the productivity of crops can be severely limited, which has a huge impact on food supplies and nutrient levels. The worst affected areas are in Africa, South America, China, and India. Adding zinc to soil in crop-growing areas is good for farmers (who benefit from higher yields) and consumers (who benefit from more zinc in their diet). The major staple crops of rice, wheat, and maize can all be affected by zinc deficiency, as can a number of fruit and vegetables.

Mankind has known of zinc since early times. Pliny the Elder, who wrote *Historia Naturalis*, mentioned an ointment that could be used to soothe and heal wounds, and scientists believe this may have contained zinc oxide. Zinc was officially identified by German chemist Andreas Marggraf in 1746, who examined calamine and obtained zinc by heating the ore with carbon. Marggraf realized that he had isolated a previously unnamed metal, although people had been smelting zinc for many centuries in Persia, China, and India.

Calamine lotion (a mix of zinc and iron oxides) and zinc treatments have been used to treat skin problems such as eczema, burns, and sunburn for many years. Parents will be familiar with—and highly grateful for—zinc oxide, which is a key ingredient in diaper rash cream.

More than 12 million tons (11 million tonnes) of zinc are produced globally each year, and the key mining areas are in Canada, Australia, the USA, and Peru. At least 55 percent of zinc is used to galvanize steel and protect it from corrosion; 17 percent is used to create alloys for die casting, and another 12 percent is required to produce bronze and brass.

Zinc may seem a highly practical element, but it has its aesthetic side, too. Rolled zinc sheeting became mandatory for roofing in Paris during the 1860s and this created the city's silvery patina, which has inspired painters ever since.

Zn 30

Zinc is a bluish white, lustrous metal. It plays multiple roles in biology, with the human body containing about 0.14 ounces (4 g) of it.

Gallium

Category: post-transition metal
Atomic number: 31

Atomic weight: 69.723
Color: silvery white
Phase: solid

Melting point: 86 °F (30 °C)
Boiling point: 3,999 °F (2,204 °C)
Crystal structure: orthorhombic

Gallium is a silvery white metal that is so soft that it can be cut with a knife. It is one of the few metals, along with mercury, cesium, and rubidium, that will melt in your hand at (or just above) room temperature, and it will also remain liquid up to 4,302 °F (2,373 °C), which has led to its use in high-temperature thermometers. It is stable in air and water, but will react with acids and alkalis.

Mendeleev predicted the existence of gallium in 1869, when he was devising his periodic table. He inferred that there was an unfilled space beneath aluminum and estimated correctly that the missing element would have an atomic weight of around 68. Gallium was actually discovered in 1875 by Paul Émile Lecoq de Boisbaudran in Paris, using spectroscopy. He was examining the spectrum obtained from a sample of zincblende ore and noticed two violet lines that he had never seen before, which he knew indicated the presence of a new element. Boisbaudran later isolated a sample of the metal and claimed it as the missing element 31. He named it gallium, supposedly for his home nation (*Gallia* is the Latin name for France), although some believe he was playing a scientific joke and naming it after himself: while *le coq* is the French for rooster, the Latin is *gallus.* Perhaps Boisbaudran was doing some crowing of his own, although he later denied it.

Gallium does not exist in its free form in nature. It can be extracted in trace amounts from the aluminum ore bauxite, although most is produced as a by-product of zinc and copper refining.

The most important uses of gallium in modern times have been in medicine and electronics. The radioactive isotope gallium-67 has a half-life of 78 hours and it will locate and gather in cancerous growths, assisting diagnosis. Gallium is also of interest in the treatment of malaria; a team of US scientists have found that a gallium compound is effective in attacking strains of malaria that have developed resistance to conventional drugs such as chloroquinine.

Gallium has also proved useful due to its semiconductor properties, in particular as gallium arsenide. Gallium semiconductors offer faster performance than more conventional silicon semiconductors and are useful for supercomputers and cell phones. Gallium is also present as various compounds in light-emitting diodes (LEDs).

Chemists love to play practical jokes and gallium is the source of one of their favorite tricks: the disappearing spoon, whereby the practical joker fashions a spoon out of gallium and gives it to unsuspecting guests, who use it to stir their tea or coffee. As they lift their spoon out, they are dumbfounded to find that most of it has disappeared into the drink!

If you can get your hands on some gallium, it's best not to handle it without some protection. It is not known to be toxic, but it will stain your skin dark brown.

The chemical element gallium is a soft, post-transitional metal. Named after the Latin word for France, it will melt at just above room temperature.

Germanium

Category: metalloid
Atomic number: 32

Atomic weight: 72.631
Color: gray–white
Phase: solid

Melting point: 1,720 °F (938 °C)
Boiling point: 5,131 °F (2,833 °C)
Crystal: diamond cubic

Germanium was discovered in 1886 by the German chemist Clemens Winkler, who decided to name it in honor of his home country. Winkler had been sent a sample of ore recovered from a silver mine at St Michaelis, near Freiberg. Winkler performed a number of chemical tests and found the ore, which had been given the name "argyrodite," to be composed of 75 percent silver and 18 percent sulfur—the remaining 7 percent was made of an unknown substance.

Fifteen years earlier, Dmitri Mendeleev—the originator of the periodic table—had suggested that there should exist an element to fill the gap between silicon and tin. He even went so far as to predict the properties of this new element. Lo and behold, Winkler's mystery substance matched Mendeleev's chemical prescription perfectly. A new element had been discovered.

The first semiconductor devices were made from germanium. These were diodes, electrical components that only allow current to pass through in one direction. This is useful, for example, in converting "alternating" electrical current (AC, of the sort that might be produced by a dynamo generator) into a steady, constant "direct current" (DC). This application of germanium was discovered by US researchers in the midst of the Second World War. The problem they faced was that the bulk of known germanium deposits were located in Germany. The eventual solution was to extract the element from the residues produced by a zinc smelting plant in Oklahoma. Although silicon (see page 40)

has since risen to dominate the field of semiconductor electronics, germanium has made something of a comeback—being used in solar panels, where sunlight falling on n-p semiconductor junctions generates useful electrical current.

Despite this, germanium's major use today is in fiber optics. Indeed, an estimated 35 percent of germanium produced is used in the cores of fiber-optic cables, where its high refractive index (which stops light escaping) and low optical dispersion (keeping the light collimated) mean it is an ideal material. Germanium is also completely transparent at infrared wavelengths, making it extremely useful in the optics of devices such as night-vision goggles, as well as infrared detectors for scientific applications.

Worldwide production of germanium is around 110 tons (100 tonnes) annually—much less than its sister element, silicon. This is a reflection of the fact that, whereas silicon is plentiful (for example, in sand and other rocky minerals), germanium makes up just 1.6 parts per million of the Earth's crust. Accordingly, germanium fetches around 80 times the price of silicon.

Ingesting limited quantities of germanium is not hazardous. Some foods—such as garlic, vegetables, and grains—contain small amounts of the element, and these aren't considered harmful. Bizarrely, however, germanium has been sold as a nutritional supplement in the USA and Japan. Studies have shown that these tablets are worse than useless—actually being harmful when taken over a prolonged period, leading to kidney problems and nerve damage.

Germanium is a hard, gray-white metalloid. It has applications in the semiconductor industry; electronic components such as computer chips are made from it. Unsurprisingly, it was discovered by a chemist from Germany.

Arsenic

Category: metalloid
Atomic number: 33

Atomic weight: 74.92160
Color: gray
Phase: solid

Melting point: n/a
Sublimation point: 2,889 °F
(1,587 °C)
Crystal structure: trigonal

It's a little-known fact that Napoleon Bonaparte—the mighty, albeit diminutive, former Emperor of France—may have been killed by his wallpaper. During the 19th century, the toxic chemical element arsenic was used as dye, in particular in a shade known as "Paris Green," popular for wallpaper. The trouble was that whenever the climate became damp and mold started to grow, the mold would ingest the arsenic in the dye and release it as gas: very bad news for anyone sat nearby. After Napoleon's defeat at the Battle of Waterloo, he was exiled to the damp island of St Helena in the south Atlantic Ocean, where he died six years later. Later analysis has revealed that Napoleon did indeed have traces of arsenic in his body. More tellingly, a scrapbook discovered in the 1980s containing—bizarrely—a sample of his wallpaper, proved it to contain the infamous Paris Green dye.

Napoleon isn't the only victim of arsenic. In Manchester, England, in the year 1900, some 6,000 people were poisoned after drinking beer that been inadvertently brewed using sugar contaminated with the element. Seventy of them died. Of course, not all arsenic deaths were accidental. Its toxic properties soon became well known and this, combined with the ready availability of arsenic sources—such as flypaper and weedkiller—led to it becoming the poisoner's substance of choice.

This was because it was, for a time, impossible to detect in the body. Just 0.0035 ounces (100 milligrams) of arsenic is enough to kill, but administering smaller doses over an extended period causes symptoms that are difficult to distinguish from ordinary illness. These include vomiting, diarrhea, dehydration, coma, and ultimately heart failure and death. Bumping off the unpopular was suddenly very easy. Ground arsenic became known as "inheritance powder" for its ability to expedite the demise of rich uncles, not to mention monarchs and popes. That all changed in 1836, when British chemist James Marsh discovered a chemical test capable of isolating the substance in biological samples.

Arsenic is an element that's long been known about. The name derives from the ancient Greek word, *arsenikon*—a name for the yellow-colored mineral "orpiment," a form of arsenic sulfide. The 5,000-year-old ice mummy "Otzi" (found in the Tirolean Alps in 1991) had traces of arsenic in his body, the likely explanation being that he was a copper smelter by trade. Copper ore is rich in arsenic. Indeed, most arsenic produced in the modern world is done so as a by-product of the extraction of other metals from their ores.

Arsenic has another modern application beyond just being a choice poison. It's used in semiconductor electronics as a "dopant," which when combined with base semiconductors converts them into n-type (electron-rich) materials for use in transistors and microchips—a prime example being gallium-arsenide.

Strangely enough, arsenic is also believed to have a beneficial effect on some lifeforms—there is evidence that chickens, for example, benefit from ingesting low doses of the element.

The chemical element of choice among poisoners through the ages: arsenic. Aside from knocking off irritating relatives and old adversaries, arsenic is used as a "dopant" in semiconductors, giving the material a net negative electric charge.

Selenium

Category: nonmetal
Atomic number: 34

Atomic weight: 78.96
Color: metallic gray
Phase: solid

Melting point: 430 °F (221 °C)
Boiling point: 1,265 °F (685 °C)
Crystal structure: hexagonal

If you know someone suffering from particularly foul breath and body odor, instead of keeping your distance, you may want to have a word in their ear about selenium, an element that can cause those side effects if taken to excess. Selenium is an essential element for humans, however, so it's important to maintain a healthy balance of it in our bodies. It forms part of the antioxidant enzyme glutathione peroxidase, which helps the body to get rid of peroxides before they can form dangerous free radicals. It is also found in another enzyme, deiodinase, which is important in hormone production in the thyroid gland.

The recommended daily intake of selenium for humans is 65–75 micrograms and anything more than 0.00017 ounces (5 milligrams) will result in symptoms of toxicity. Deficiency in selenium is more likely to occur, however, particularly in some parts of the world where the soil is lacking in this element. Good dietary sources of selenium include bread, breakfast cereal, oily fish, nuts, and bran.

Interestingly, selenium intake fell during the late 20th century due to the decrease in popularity of foods such as liver and kidneys, which are rich sources. Some medical researchers have highlighted a link between decreasing selenium intake and both increasing cancer levels and declining sperm counts. Controlled clinical tests in the 1990s showed that men with low sperm counts who were given selenium showed a marked improvement in their condition, while sperms counts in those given a placebo remained the same.

In some areas, selenium is added to soil to maintain good health in livestock. However, if selenium levels are too high this can be toxic to animals and cause a condition known as "blind staggers"; the affected animal walks with an unsteady, staggering gait and appears to be blind. The plant milk vetch (*Astragalus*) can absorb dangerous quantities of selenium and is particularly palatable to cattle and horses. Soil of the Great Plains of North America is rich in selenium and cowboys used to call vetch "locoweed," from the Spanish word *loco*, meaning "insane." Scientists have observed that if other sources of selenium were to become low, the element could be harvested from ordinary soils by growing plants such as milk vetch that would absorb it for later collection.

Selenium is a member of group 16 of the periodic table, and is positioned below sulfur and tellurium. It was discovered in 1817 by Jöns Jakob Berzelius, a professor of chemistry and medicine who worked at a sulfuric acid production plant in Sweden for a few months that year. He was particularly interested in the brown–black sediment left in the bottom of the chambers in which the acid was made. At first, Berzelius thought it was tellurium (see page 124) but although the sediment was chemically similar to this element, he soon realized that he had found something new. Berzelius named his discovery selenium (from the Greek word *selene*, meaning "Moon") to correspond with tellurium (from the Latin word *tellus*, meaning "Earth").

Despite its shiny appearance, selenium is, in fact, a nonmetal. While it is believed to be an important dietary mineral for good health, too much of it will give you embarrassing halitosis and body odor.

Bromine

Category: halogen
Atomic number: 35

Atomic weight: 79.904
Color: deep red
Phase: liquid

Melting point: 19 °F (-7 °C)
Boiling point: 138 °F (59 °C)
Crystal structure: orthorhombic

The name bromine is derived from the Greek word *bromos*, which means "stench." Taking a whiff of this reddish liquid reveals a smell similar to that of chlorine, but don't try this at home; in its natural state, bromine is highly toxic and reactive—like most of its fellow halogens.

Bromine rarely occurs in its natural state, but bromide (a bromine atom that has acquired a negative charge) is stable, safe, and plentiful. Its solubility has resulted in its accumulation in the oceans, mineral water springs, and brine residues. The element was discovered by two chemists working independently. Antoine-Jérôme Balard found bromine in residues from the salt marshes of Montpellier, France. By passing chlorine gas through them, he was able to isolate a reddish liquid that he deduced was a previously unknown element. He sent a report of his findings to the journal of the French Académie des Sciences in 1826, and academy members named the liquid bromine. A year earlier, student Carl Löwig isolated bromine from a spring near his home at Kreuznach, Germany, which he took to his professor at the University of Heidelberg. As Balard, however, published his results prior to Löwig, it was Balard who took the credit for the element's discovery.

Bromine is still produced by bubbling chlorine gas through seawater, which converts dissolved salts to bromine. Air is blown through the water and the bromine gas is expelled and collected.

People have been using bromine compounds since ancient times. The togas worn by Roman emperors were colored Tyrian purple using a dye that contained bromine atoms. This was developed from a mucus secreted by the Mediterranean mollusk *Murex brandaris* and became highly prized in ancient cultures.

In the 19th century, bromide salts were often prescribed by doctors as sedatives and anticonvulsants for people who had epilepsy. Bromide also has an anaphrodisiac effect (it suppresses libido). It is rumored that, during the First World War, the British Army laced soldiers' tea with potassium bromide in order to quell sexual appetite, and French soldiers were also said to be given wine adulterated with bromide. It seems unlikely, however, that military leaders would administer a sedative to troops waiting to go into battle.

In modern times, bromine has been used in organobromo compounds—which contain carbon bonded to bromine. These are used extensively in fire extinguishers and as flame retardants in furniture foams and textiles. When the material burns, the flame retardant produces an acid that disrupts the normal oxidation process of the fire. Doubts have been raised about organobromo compounds, however, as a result of their effect on the environment. Methyl bromide (CH_3Br) was used as a soil fumigant from the 1950s to kill insects, nematodes, bacteria, and fungi. It was particularly effective because the gas could penetrate areas of the soil that other pesticides failed to reach and pests did not develop resistance to it. However, in 1992 it was added to the list of ozone-depleting substances by the Montreal Protocol and began to be phased out.

Bromine vapor (orange) diffusing upward. It is used in fire retardants and for pest control—though, contrary to popular stories, probably not for quelling soldiers' libidos.

Krypton

Category: noble gas
Atomic number: 36

Atomic weight: 83.798
Color: colorless
Phase: gas

Melting point: -251 °F (-157 °C)
Boiling point: -244 °F (-153 °C)
Crystal structure: n/a

If you just happen to be reading this book on a plane that is about to come into land at night, be grateful for the element krypton. Many of the high-powered, flashing lights that help guide the pilot toward the runway will be lit by this fascinating element.

Although Earth has retained all of the krypton that was present when it was formed around 4.5 billion years ago, krypton is still one of the rarest gases in our planet's atmosphere, constituting around one part per million of air. Like its fellow members of the noble (or "inert") group of gases, it's colorless, odorless, and tends not to react chemically with other elements, although it will bond with fluorine (see page 30). Krypton is extracted by a distillation process from liquid air—air that has been cooled to -319 °F (-195 °C). Solidified krypton is crystalline and white, with a face-centered cubic crystal structure.

The element was discovered in 1898 by the British chemist Sir William Ramsay, who was based at University College, London, England. Having already isolated helium and discovered argon, Ramsay saw a new group of elements emerging that shared similar characteristics; he was convinced that he could discover more. With his assistant Morris Travers, Ramsay began an in-depth study of argon gas, which he had extracted from air by chemically removing the other gases present: nitrogen, oxygen, and carbon dioxide. They liquefied the argon and allowed it to evaporate, to see if any other elements remained.

Their efforts were successful; from the 26 pints (15 liters) of argon that they began

with, Ramsay and Travers were able to isolate 0.8 fluid ounces (25 ml) of another gas. They tested it in an atomic spectrometer and noted orange and green lines that could not be attributed to any other known gas. The pair named their find krypton, from the Greek word *kryptos*, meaning "hidden," and went on to discover neon and xenon. Ramsay was awarded the Nobel Prize in Chemistry in 1904 for his contribution to the discovery of the inert (or "noble") group of gases.

It's intriguing to note that Mendeleev, who died in 1907, did not predict the existence of this new group of gases on the periodic table. Although this disappointed him, he accepted Ramsay's discovery as being further proof of the table's validity.

Like other noble gases, krypton lights up when an electric current is passed through it, producing a distinctive, eye-catching blue-white glow. It can also be mixed with other gases to create a bright, greenish-yellow tone. Krypton has been used in high-speed photographic flash lights and strobe lighting, because it has a particularly swift response to electric current. Also, krypton is one of the gases used in the krypton fluoride laser, a highly intense, deep ultraviolet laser that has been important in surgery, the manufacture of microelectronic devices and circuits, and nuclear fusion energy research—where the laser is used to heat a pellet of fusion fuel to start a nuclear reaction.

Between 1960 and 1983, one of krypton's spectral lines played an important role in defining the meter as a unit of length—it was specified as "1,650,763.73 wavelengths

A krypton-gas discharge lamp. The sample of gas is contained in a borosilicate vial, around which are wrapped silver-coated copper wires. An electric current is passed through the wires, ionizing the gas in the vial—turning it into a "plasma" and making it glow. Krypton gives off a characteristic blue-white glow.

of the orange–red emission line in the electromagnetic spectrum of the krypton-86 atom in a vacuum." This was later replaced with a definition based on the distance traveled by light in a vacuum in a specified interval of time.

Krypton has become useful in energy-saving fluorescent lighting and high-efficiency lightbulbs. The old style of energy-inefficient incandescent bulbs were usually filled with nitrogen or argon; krypton, which has a higher number of molecules per unit volume than either of these elements, reduces the evaporation of the tungsten filament, keeps the bulb working longer at high temperatures, and ensures that more energy goes into creating light than heat. Even though most people are being encouraged to replace incandescent lights with fluorescent bulbs, this use of krypton in the older-style bulb does make it more efficient than it was. LEDs (light-emitting diodes) are still the most energy-efficient lightbulbs, but they are an expensive option.

Krypton has a couple of particularly interesting isotopes. Scientists are using Kr-81 to date old groundwater, which collects beneath Earth's surface after filtering down through sand and rock fissures. Many people across the world depend on groundwater for drinking and good harvests. Therefore, scientists needed an accurate way to date the water or to see how long it had been underground—in order to gauge whether it might be contaminated. They have been able to do this by using Kr-81, which is present in tiny quantities in air. Water picks up traces of Kr-81 when it is above ground, which it retains when it drains into the earth. This krypton isotope has a half-life of 230,000 years, so scientists can gauge how old the groundwater is by counting the remaining Kr-81 atoms in a sample. Some groundwater can be dated back 800,000 years.

During the Cold War, the isotope Kr-85 was used as a "chemical spy." It is produced in nuclear reactors and reprocessing plants, from where some of this radioactive isotope may escape into the atmosphere. Kr-85 levels were monitored by the West to help gauge the nuclear activity of Eastern Bloc countries. By comparing these readings with the amount of Kr-85 produced by reactors in the USA and Europe, it was possible to calculate the scale of the nuclear operation taking place behind the Iron Curtain. This gave an indication of the reserves of nuclear material Eastern Bloc countries had available to produce weapons.

On a lighter note, if you remember the Superman comics or have been a fan of the eponymous movies, you may be interested to note that the mineral *kryptonite* is not just the stuff of superhero fiction. Born on the planet Krypton, Superman's legendary powers are sapped on exposure to kryptonite's green crystals. In 2007, scientists found a mineral in Serbia that matched the exact chemical formula of kryptonite as explained in the movie *Superman Returns* (2006). The mineral, known as jadarite, is white instead of green and doesn't glow in daylight, although it does react to ultraviolet light by producing a pink–orange fluorescent glow.

Ionized krypton gas is sometimes used in lasers, such as the one pictured here. An electric current "pumps" the gas full of energy, which is then released in the form of laser light—all the same wavelength (or color) and focused into a narrow beam.

Rubidium

Category: alkali metal
Atomic number: 37

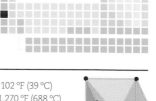

Atomic weight: 85.4678
Phase: solid
Color: silvery white

Melting point: 102 °F (39 °C)
Boiling point: 1,270 °F (688 °C)
Crystal structure: body-centered cubic

Rubidium is a soft, silvery metallic element belonging to group 1 of the periodic table: the alkali metals. It shares a number of their characteristics in that it's highly reactive and oxidizes rapidly in air. Rubidium reacts violently with water and is so volatile that it must be stored under grease or oil to stop it from igniting spontaneously in air.

The element was discovered in 1861 by the chemists Robert Bunsen (of Bunsen burner fame) and Gustav Kirchhoff of the University of Heidelberg in Germany. The duo knew that lepidolite, a mineral discovered by Jesuit priest Nicolaus Poda in the 18th century, contained lithium and they believed that it might contain another alkali metal as well.

As with a number of elements, rubidium was detected through a process called spectroscopy. This involves heating the element, which creates a spectrum of lines that can be analyzed; each element will create a characteristic spectrum, indicating its presence. The two scientists marveled at the intense ruby red rays of their find and decided to name the element after the Latin word *rubidius*, meaning "deepest red."

There is no known biological role for rubidium, but the element is so similar to potassium that it is easily absorbed by the human body and plants. Some plants, if they are particularly deficient in potassium, will absorb rubidium and appear to thrive.

Rubidium is used in atomic clocks, which are the most accurate time devices (see also cesium, page 130). Electrons circle the nuclei of atoms in orbits, and each orbit has an extremely well-defined energy. If an electron absorbs energy, it can jump up to a higher energy level, dropping back to its original orbit a short time later. As it drops back, it emits a burst of radiation with energy exactly equal to the energy gap between the two orbits. Because the energy of radiation is directly linked to its frequency, this means that each atom gives off radiation with a very sharply defined frequency. In the case of rubidium this is 6,834,682,610.904324 cycles per second and the clock uses this well-defined frequency rather like a superfast pendulum. Each tick counts off one-6,834,682,610.904324th of a second, and that's what makes atomic clocks so accurate.

There are two naturally occurring isotopes of rubidium—the stable Rb-85, which accounts for 72 percent of the element on Earth, and radioactive Rb-87, which accounts for the remaining 28 percent. Rb-87 has a half-life of around 50 billion years, which is three times the age of the universe.

Rubidium-strontium dating is a method of estimating the age of rocks. This is done by measuring the amount of the isotope strontium-87 that has formed due to the decay of rubidium-87, which was present when the rock was formed. This method has been particularly useful in dating very old rocks, because the transformation is slow (due to the long half-life of Rb-87). This isotope may be considered weakly radioactive, but rubidium contributes to the background radiation on Earth and a sample will produce sufficient radiation to blacken a photographic film in 110 days.

Rb
37

Like its fellow group members potassium and
cesium, and to the delight of school children the
world over, Rubidium is another chemical element
that reacts violently when dropped into water.
Because of its reactivity, rubidium is stored in
glass vials filled with an inert gas such as argon.

Strontium

Category: alkaline earth metal
Atomic number: 38

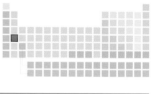

Atomic weight: 87.62
Color: silver–gray
Phase: solid

Melting point: 1,431 °F (777 °C)
Boiling point: 2,520 °F (1,382 °C)
Crystal structure: face-centered cubic

Strontium is a soft, silvery, highly reactive alkaline earth metal. It is named after Strontian, the Scottish village where it was first discovered in ores taken from nearby lead mines. In 1790, local physician Adair Crawford recognized a new mineral (strontia or strontianite) in witherite (a mineral containing barium carbonate) and published a paper on his findings. This attracted the attention of chemist Thomas Charles Hope, who proved that strontianite contained a previously undiscovered element and noted that it made flames burn red.

In 1799, another strontium mineral was found in Gloucestershire, England, where it was being used for decorative gravel paths in backyards and gardens. This mineral contained strontium sulfate and became known as celestite. In 1808, Sir Humphry Davy was the first person to isolate strontium by electrolysis of strontium chloride and mercury oxide—the same process he had used to isolate barium and calcium.

If added to water, strontium reacts to produce strontium hydroxide and hydrogen. If exposed to air, strontium metal quickly turns a yellowish color with the formation of the oxide, while finely powdered strontium metal will ignite spontaneously in air. Strontium carbonate and other strontium salts form the bright red color in fireworks and warning flares.

Strontium is often associated with its radioactive isotope strontium-90, which was produced by above-ground nuclear explosions during the mid-20th century and became a major source of contamination. Strontium-90 is a particularly powerful emitter of harmful radiation and has a half-life of 29 years. A study of hundreds of thousands of teeth in the 1950s and 1960s showed that levels of strontium-90 were 50 times higher in children born in 1963 in than those born in 1950, prior to above-ground nuclear testing. These findings helped to convince President John F. Kennedy to sign the Partial Nuclear Test Ban Treaty with the United Kingdom and Russia, ending above-ground nuclear weapons testing. The Chernobyl disaster of 1986, when an explosion at a nuclear power plant in the Ukraine spread radioactive fallout over western Russia and most of Europe, contaminated a huge area with strontium-90.

Ordinary strontium is not radioactive, but it has certainly been tainted by association with its dangerous isotope. Perhaps that is why it has few other uses in general. Strontium compounds are used in glass for old-fashioned color television cathode ray tubes to prevent X-ray emission, and strontium metal can be used as a "getter" to remove the last traces of air from vacuum tubes. Strontium chloride is sometimes used in toothpastes for sensitive teeth; it forms a barrier over areas of the tooth that have been exposed by gum recession.

The human body absorbs strontium as though it were calcium and the two elements are very similar chemically. Dietary supplements containing strontium are sold as "bonemakers," and a study carried out by the New York College of Dental Sciences using strontium on osteoblasts (bone cells) showed a marked improvement in growth.

A sample of strontium. The radioactive isotope strontium-90 is a component in the fallout from atom bomb explosions. Strontium is chemically similar to calcium, which means that the radioactive isotope can be absorbed by bone and lead to serious health problems, including cancer.

Yttrium

Category: transition metal
Atomic number: 39

Atomic weight: 88.90585
Color: silver–white
Phase: solid

Melting point: 2,779 °F (1,526 °C)
Boiling point: 6,037 °F (3,336 °C)
Crystal structure: hexagonal

Yttrium was first noticed in 1787 near the village of Ytterby in Sweden, by an army lieutenant and amateur chemist named Carl Axel Arrhenius. Arrhenius found a lump of black rock, resembling coal, in a quarry close to the village. Perplexed by its unusually heavy weight, he sent the rock to chemist colleagues for further analysis. The rock was found to contain a new mineral (which later became known as "gadolinite," after one of the chemists). Later analysis in 1843, by Swedish chemist Carl Gustaf Mosander, revealed gadolinite to harbor the oxide of a new element—called yttrium, in honor of the place of its discovery. In fact, the mineral was soon found to also contain oxides of two other new elements—erbium and terbium. Amazingly, 35 years later, a fourth element emerged from gadolinite, which was named ytterbium.

Yttrium is sometimes called a "rare earth" element because it is very similar in form and properties to the lanthanide rare earth series (see page 134)—and tends to be found in the same deposits of ore as the lanthanides. It is never seen naturally in its pure metallic form. In fact, pure yttrium metal wasn't extracted from raw minerals until 1828, when German chemist Friedrich Wohler combined yttrium chloride with potassium to form potassium chloride plus the elusive silver white metal.

Yttrium has a host of important applications. Over 30 isotopes of yttrium have been observed, all of which are radioactive. One of these, Y-90, a product of strontium decay, is particularly effective in cancer treatment. The chemistry of this isotope enables it to bind with antibodies that seek out and bind themselves to cancer cells, enabling the yttrium's radiation to selectively destroy them.

Yttrium is used extensively in the production of synthetic garnets. In particular, yttrium aluminum garnets (YAGs—chemical formula $Y_3Al_2(AlO_4)_3$) form very high-quality crystals. This led to their use as fake diamonds in jewelry before cubic zirconia emerged as a superior alternative. YAGs are still used today in lasers—the YAG crystal is pumped with energy, which it then reemits as a laser light. YAG lasers are extremely powerful, capable of cutting through metal, reliable enough to be used as a surgical tool, and bright enough to be bounced to the Moon and back in experiments to establish the distance to our satellite.

Perhaps the greatest application of yttrium, however, is in the development of superconductors. Most electrical conductors exhibit a phenomenon called "resistance"— which impedes the flow of electric current. Superconductors have zero resistance, meaning a current could—in theory— continue to circle in a coil indefinitely. The trouble is that superconductors generally have to be chilled to near absolute zero, -460 °F (-273 °C), using cumbersome cooling systems. In 1987, however, it was discovered that the compound yttrium barium copper oxide (YBCO) becomes a superconductor at -292 °F (-180 °C)—still cold, but warm enough so that liquid nitrogen can be used as a coolant rather than the more expensive (and colder) liquid helium.

The rare transition metal yttrium. An alloy of yttrium with barium and copper oxide becomes a superconductor at -292 °F (-180 °C)—a perfect conductor of electricity, with zero electrical resistance.

Zirconium

Category: transition metal
Atomic number: 40

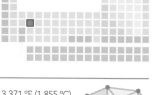

Atomic weight: 91.224
Color: silvery white
Phase: solid

Melting point: 3,371 °F (1,855 °C)
Boiling point: 7,968 °F (4,409 °C)
Crystal structure: hexagonal close-packed

Zirconium was discovered in 1789, by German chemist Martin Heinrich Klaproth, who was investigating "jargon" (a gemstone found in Sri Lanka). This analysis led to the production of an oxide, which he called zirconia. The name came from "zircon," a yellow-colored mineral related to jargon—named after the Arabic word *zargun*, which means "gold colored." Klaproth's studies revealed that zirconia must be formed from a new element, which he named zirconium.

Although Klaproth inferred its presence, pure zirconium metal was not produced until 1824, by Swedish chemist Jöns Jakob Berzelius (see also selenium, page 84). It's a soft, silver–gray substance, often added to other metals to form alloys with enhanced workability. It constitutes around 0.00208 ounces per pound (130 mg per kg) of the Earth's crust.

Zirconium is probably best known to most people for zirconia—zirconium dioxide, the substance that Klaproth discovered first. The face-centered cubic crystal allotrope of zirconia, known as "cubic zirconia," has found widespread applications as a replica gemstone—resembling, to all but the highly trained eye, a high-quality diamond. It's even been argued by one commentator that cubic zirconia isn't cheap diamond but, conversely, diamond is an expensive alternative to cubic zirconia!

But zirconium has more going for it than superficial glamor. It's used as a piping material inside nuclear reactors. Zirconium is the metal of choice for such applications because it doesn't interact at all with neutrons. Neutrons are the particles given off by nuclear fission reactions and that trigger subsequent reactions—thus sustaining the nuclear chain reaction. Materials that absorb neutrons would interfere with the pace of the reaction.

Zirconium's light weight and its resistance to heat—not melting until 7,968 °F (4,409 °C)—have led to applications in spacecraft, as well as in jet engines and abrasive materials. It's also been used in lab equipment for high-temperature chemical reactions, and to make tools and fittings for foundries—such as ladles, furnace linings, and molds—that are capable of withstanding the searing heat needed to melt metals such as iron and steel.

Like titanium, a fellow group 4 member in the periodic table, zirconium is highly resistant to corrosion, forming a thin skin of oxide over its surface upon contact with the air, which then protects it against further oxidization.

Zirconium's oxide zirconia has also been developed into a ceramic material. The first example of this was a type of ceramic that was strong enough to use to build an engine. As it wasn't built from metal, the engine didn't need lubricating with oil, nor did it require any cooling—which meant that it soon found a military application powering tanks. The same zirconia-based ceramic is also used in ultrasharp knife blades: zirconium dioxide powder is pressed into a mold and fired at high temperature, producing a blade that (once sharpened on a diamond grinder) rarely needs resharpening. The technique has now been seconded to make everything from ceramic golf clubs to dental crowns to turbine blades.

Zirconium is not much to look at. However, the derivative compound zirconium dioxide, also known as "zirconia," has a cubic crystalline form that is both transparent and sparkly, leading jewelers to use it as a substitute for diamond.

Zr 40

Niobium

Category: transition metal
Atomic number: 41

Atomic weight: 92.90638
Color: steel gray
Phase: solid

Melting point: 4,491 °F (2,477 °C)
Boiling point: 8,571 °F (4,744 °C)
Crystal structure: body-centered cubic

Used in jewelry and body piercings, niobium rather suits its exotic name, which is derived from Niobe, daughter of King Tantalus in Greek mythology. In the same way, niobium has a family connection to tantalum: they too are very similar chemically, both being transition metals in group 5. Niobe was known as the goddess of grief after all 12 of her children were murdered.

Niobium is a soft, shiny, gray metal when pure and has no known biological role. It was discovered in 1801 by the British chemist Charles Hatchett, who was examining a sample of the mineral columbite sent to the British Museum from Connecticut in the USA. Hatchett heated the columbite with potassium carbonate and dissolved the product in water. He then added acid to form an oxide powder. This precipitate intrigued Hatchett, who deduced that it must harbor an unknown metal.

Hatchett named the element columbium, based on a poetic name for America, but debate continued for many years over whether or not columbium was another form of tantalum, which was discovered the following year. In 1844, German chemist Heinrich Rose proved that columbite contained both tantalum and columbium—he then renamed the latter element niobium.

In 1864, chemist CS Blomstrand produced a sample of pure niobium metal by heating niobium chloride with hydrogen. For many years, the element was extracted mainly from columbite and as the mineral

pyrochlore, which has become the main source today. The main mining areas are Brazil (which produces 85 percent of niobium), Russia, Nigeria, Zaire, and Canada. World production is estimated at around 27,500 tons (25,000 tonnes) per year.

Most niobium is used in the production of high-grade steel, which has improved strength and malleability. With nickel, cobalt, and iron, niobium is used to make superalloys for applications such as jet-engine components, gas turbines, rocket components, and other heat-resistant equipment. Niobium-germanium, niobium-tin, and niobium-titanium alloys are used to make the superconducting magnets used in magnetic resonance imaging (MRI) equipment and in particle accelerators. The Large Hadron Collider in Geneva uses 661 tons (600 tonnes) of superconducting wire.

Like titanium, tantalum, and aluminum, niobium can be heated and anodized (a process that increases the thickness of the oxide layer on the surface of metal, thus boosting its resistance to corrosion). This creates myriad beautiful colors as light is passed through the oxide layers. Niobium has also been used to create commemorative coins, often with gold or silver.

As niobium is hypoallergenic as well as resistant to corrosion, it's a good choice for jewelry and body piercings. Alloys of niobium can be used for surgical instruments and medical devices such as pacemakers, because they do not react with human tissue.

Niobium may not be a household name, but if you have a pacemaker fitted, or have ever had an MRI scan to detect medical conditions, you owe a lot to this intriguing metal.

Molybdenum

Category: transition metal
Atomic number: 42

Atomic weight: 95.94
Color: silvery white
Phase: solid

Melting point: 4,753 °F (2,623 °C)
Boiling point: 8,382 °F (4,639 °C)
Crystal structure: body-centered cubic

Molybdenum must be one of the most challenging elements of the periodic table to pronounce correctly, so here's its phonetic spelling: mol-ib-den-uhm. The name itself is derived from the Greek word *molybdos*, meaning "lead"; this is because its ores were at one time confused with lead ore.

Molybdenum does not occur naturally in its pure form on Earth and must be obtained from ore, the principle one being molybdenite, which was originally called molybdena. Molybdenite and graphite look very similar, which is why the two were often confused and were both used to make pencils. In 1778, the Swedish chemist Carl Scheele (see also chlorine, page 48) realized that molybdenite was neither lead nor graphite, and he and other chemists deduced correctly that it contained a previously unknown element. In around 1781, Scheele's friend Peter Jacob Hjelm, of Uppsala, Sweden, successfully isolated the pure element.

Today, molybdenum is used mainly to create particularly tough steel alloys, which have great strength and heat resistance. During the First World War, demand for molybdenum increased due to its use in tanks. The British forces were the first to deploy tanks on the Western Front and, although early models were armor-plated with 3-inch (75-mm)-thick manganese-steel plates, they were still not strong enough to resist direct hits. A 1-inch (25-mm) layer of molybdenum steel was introduced, which not only gave greater protection, but, due to the thinner layer being much lighter, also improved the tank's maneuvrability and speed.

Today, molybdenum steel (often referred to as "moly steel") is a workhorse of industry, used in areas where applications or parts need to withstand high temperatures (for instance, in the automobile and aircraft industries and for rocket engines). It also has good electrical conductivity and resistance to corrosion and wear. Molybdenum alloys are used for high speed tools such as drills and saw blades, and in heating elements.

Molybdenum is an essential element to all species, and is a component in a range of biological enzymes. It is not known how much molybdenum human beings require each day, but it is thought to be as low as 0.05 milligrams. Good sources include pork, lamb, eggs, green beans, and soybeans. One significant enzyme containing molybdenum is nitrogenase, which is found in the root nodules of legumes and which allows the plant to convert nitrogen gas into nitrates, needed for growth.

Molybdenum and its neighbor on the periodic table, technetium, are both used in medicine. The radioactive isotope molybdenum-99 (half-life of 66 hours) decays to generate technetium-99m, which has a half-life of only six hours and is used in hospitals for medical imaging. The technetium isotope collects in various parts of the body, in particular bone. Scanners detect the radiation from the isotope to create an image, enabling diagnosis. The process uses a device that is filled with Mo-99, which decays into Tc-99m, forming a continuous supply that is "milked off." The device is nicknamed the "moly cow."

A sample of pure molybdenum foil. Mention molybdenum to most people and you'll get blank stares in return. It is, however, all around us in the form of "moly steel," a construction-grade steel alloy that is used for skyscrapers, huge bridges, and rocket engines.

Technetium

Category: transition metal
Atomic number: 43

Atomic weight: 98
Color: silvery gray
Phase: solid

Melting point: 3,915 °F (2,157 °C)
Boiling point: 7,709 °F (4,265 °C)
Crystal structure: hexagonal

Technetium (counterintuitively pronounced tek-nee-shee-um) is a radioactive metal that is silvery in its metallic form and is often obtained as a gray powder. It is found in trace amounts in uranium ore and most of it is produced inside nuclear reactors.

The element reveals important aspects of astrophysics. The longest living isotope of technetium has a half-life of four million years, which means that any technetium that existed when Earth was formed 4.57 billion years ago would be long gone. In 1952, astronomer Paul Merrill was examining the light given off by red giants (stars that are approaching the end of their lives) and these spectral lines indicated that the stars were rich in technetium. Given that the isotopes of technetium are relatively short lived, and that red giants are stars close to their demise and thus billions of years old, this meant that the element must be formed within the star itself. This proved to scientists that all elements heavier than lithium are created by the nuclear reactions at the heart of the stars.

There is no biological role for technetium, but it serves an important function in medical imaging. The isotope technetium-99m will bond with cancer cells and can be detected because it emits gamma-rays. The technique is known as immunoscintigraphy and is used to study the brain, lungs, digestive system, and bones. If it is combined with a tin compound, technetium-99m can bind to red blood cells to indicate circulatory problems. It can also be used to highlight the extent of the damage caused by a heart attack.

The name technetium is derived from the Greek word *tekhnetos*, which means "artificial," and this refers to the fact that it was the first nonnaturally occurring element to be created. It only exists on earth through technology, apart from those trace amounts in uranium ore.

Early forms of the periodic table contained a gap between element 42 (molybdenum) and element 44 (ruthenium). Mendeleev predicted that the "missing" element would have qualities similar to those of manganese, at the top of group 7, and named it eka-manganese (*eka* is the Sanskrit word for "one"). Its chemistry, is in fact, more similar to that of rhenium, which occupies the space below technetium.

Teams of chemists searched for the element in the late 19th and early 20th centuries and many claims were made, none of which were accepted. That was until 1937, when Emilio Segrè and Carlo Perrier at the University of Palermo, Italy, separated technetium from a sample of molybdenum that had been bombarded by deuterons in a cyclotron at the University of California, Berkeley. In 1925, German chemists Ida Tacke, Walter Noddack, and Otto Berg had bombarded columbite mineral with a beam of electrons and detected the presence of the elusive element 43 by the faint X-rays emitted. Other chemists could not replicate their findings, so they were discounted.

Medical technetium is extracted from the spent fuel rods used to power a nuclear reactor. The isotope Tc-99 has a half-life of 200,000 years and can pass through the food chain, causing cancer in humans.

The isotope technetium-99m (where the "m" stands for metastable) is used in medical imaging to detect cancerous growths, blood disorders, and the degree of damage caused by a cardiac arrest. This image from a technetium bone scan, shows a cancer in the lower ribs.

Ruthenium

Category: transition metal
Atomic number: 44

Atomic weight: 101.07
Color: silvery white
Phase: solid

Melting point: 4,233 °F (2,334 °C)
Boiling point: 7,502 °F (4,150 °C)
Crystal structure: hexagonal

Ruthenium is one of the rarest metals found on Earth. It is in group 8 of the periodic table and is the first of the precious metals that are also classed as belonging in the "platinum group"; this is because they occur with platinum in ores and share some of its properties. Ruthenium has been found in its free state and in some minerals, the most common being laurite (ruthenium sulfide). Its minerals are rare, however, and most ruthenium is produced as a by-product of nickel refining.

The metal itself is hard, silvery white and is unaffected by air, water, and acids, although it will dissolve in molten alkalis. Due to its high resistance to corrosion, ruthenium is used as a thin plating in jewelry to give a darker, pewter-like shine. When used to cover a fairly cheap metal, this can be more cost effective than using pewter itself.

The element ruthenium was discovered over time during the early part of the 19th century. In 1807, the Polish chemist Jedrzej Andrei Sniadecki investigated platinum ores, hoping to follow the success of Wollaston and Tennant who had discovered rhodium, palladium, osmium, and iridium in this source. The following year, Sniadecki claimed that he had indeed found a new metal, which he named vestium, after the asteroid Vesta, which had been first noted a year earlier. Other chemists tried to repeat Sniadecki's work without success so he dropped his claims to the discovery. In 1825, the Swedish chemist Jöns Jakob Berzelius in Stockholm and the German scientist GW Osann at the University of Dorpat (now Tartu) separately examined platinum ore from the Ural Mountains of Russia. Berzelius could not find any new elements, but Osann claimed he had found three new metals, which he named pluranium, polinium, and ruthenium.

The first two metals were never verified, but the third claim was validated by Karl Karlovich Klaus at the University of Kazan. He highlighted that Osann's sample of ruthenium was impure and managed to extract his own pure sample. As a result, Klaus was given the credit for ruthenium's discovery, although he kept Osann's choice of name: ruthenium refers to a geographical region that existed during the middle ages and included modern-day Russia, the Ukraine, and Belarus.

There are seven naturally occurring ruthenium isotopes, none of which are radioactive. Another 34 isotopes have been discovered or produced, including the radioactive isotope Ru-106, which is produced in nuclear reactors and has a half-life of 372 days. This isotope has been found in the edible seaweed porphyra in the Irish Sea, believed to have come from discharges from the Sellafield nuclear reprocessing plant in England. Porphyra is used to make the Welsh dish laver bread, and as a result some of this bread has been contaminated with Ru-106.

The element's main uses are in catalysts and superalloys. It is a very effective hardener for platinum and palladium, and can be alloyed with these metals to make electrical contacts. It has also been added to titanium deep-water pipes to improve their resistance to corrosion.

Ruthenium is a hard, white transition metal which is extremely rare on Earth. It was originally separated from ore found in the Ural mountains of Russia, from which this element's name is derived.

Rhodium

Category: transition metal
Atomic number: 45

Atomic weight: 102.90550
Color: silvery white
Phase: solid

Melting point: 3,567 °F (1,964 °C)
Boiling point: 6,683 °F (3,695 °C)
Crystal structure: face-centered cubic

Rhodium is a beautiful, silvery white metal in group 9 of the periodic table. Its luster and reflectivity make it particularly useful in jewelry, as a thin layer of rhodium can add such shine to the metal to make it look like silver or platinum. It is corrosion resistant, unaffected by air and water up to 1,112 °F (600 °C) and impervious to attack by acids, including *aqua regia* (a highly potent mixture of acids) up to 212 °F (100 °C). Only molten alkalis can attack this precious metal.

Rhodium was discovered in 1803 by the esteemed British chemist William Hyde Wollaston, who dissolved a sample of platinum ore in acid from which he precipitated platinum then palladium. A rose-colored solid was left, from which he obtained crystals of rhodium chloride, and then reduced to the metal by heating with hydrogen. The element was named after the Greek word *rhodon*, meaning "rose-colored."

Rhodium is classed as a precious metal due to its rarity on Earth; it is 1,000 times rarer than gold. It has been found in pure form in deposits in the USA and also in the rare mineral rhodplumsite and copper and nickel ores. Principal sources are South Africa, Russia, and Canada, with world production estimated at 27.5 tons (25 tonnes) yearly.

Its rarity has resulted in rhodium's use in honors when even platinum, gold, and silver are not considered precious enough. Fans of music trivia will recall that in 1979 the *Guinness Book of World Records* presented Paul McCartney with a rhodium-plated disk, to celebrate his status as history's bestselling recording artist and songwriter.

The price of rhodium has fluctuated dramatically in recent times. Between 2004 and 2008, the value of 1 pound (450 g) of rhodium increased from US$5,000 to US$110,000. However, in 2008 the economic slowdown resulted in the price plummeting.

The main use of rhodium is in catalytic converters, which change harmful emissions from the exhausts of vehicles into less-polluting substances. Most vehicles are fitted with a three-way converter, which deals with the main offenders. Oxidation plays a dual role, converting carbon monoxide to carbon dioxide and unburned hydrocarbons to carbon dioxide and water. Reduction converts nitrogen oxides to nitrogen and water. Catalytic converters are also used on trucks, buses, airplanes, mining equipment, and locomotives. Platinum can be used as a catalyst, but it is costly and causes other unwanted reactions; therefore rhodium is more generally used as a reduction catalyst for nitric oxides, where it is particularly effective, and palladium is used as an oxidation catalyst.

Other uses of rhodium stem from its reflectivity as a thin surface layer. It has been used as a coating for optical fibers and mirrors, and in the reflectors of car headlights.

One of rhodium's most intriguing characteristics is its peculiar ability to absorb oxygen from the atmosphere without becoming oxidized itself. This is because, when rhodium is melted, it captures oxygen and then releases it as it solidifies.

Rhodium is actually rarer on Earth than gold. Used to add luster to jewelry, it also has practical uses in catalytic converters—devices for making car exhaust fumes cleaner.

Palladium

Category: transition metal
Atomic number: 46

Atomic weight: 106.42
Color: silvery white
Phase: solid

Melting point: 2,831 °F (1,555 °C)
Boiling point: 5,365 °F (2,963 °C)
Crystal structure: face-centered cubic

Palladium is a lustrous, silvery white metal, which is so malleable that, like gold, it can be beaten into leaves as thin as a micrometer. It is in group 10 of the periodic table and is also classed as a "platinum metal." Palladium is notable for being the least dense of this latter group and having the lowest melting point. It has good corrosion resistance, but dissolves in some acids and molten alkalis.

The element was discovered in platinum ore by William Hyde Wollaston and his colleague Smithson Tennant, in London in 1802. Like other chemists, Wollaston and Tennant knew that if platinum ore was dissolved in *aqua regia* (a particularly strong mix of acids), a black residue would remain, which many believed to be only graphite. The British duo thought differently and began a series of experiments that were to prove fruitful. Tennant became the first chemist to isolate iridium and osmium from the residue, while Wollaston discovered palladium and rhodium by carrying out a number of treatments on the solution. He chose the name palladium in honor of the asteroid Pallas, which was discovered at roughly the same time as the element.

Like rhodium, palladium has been of recent importance due to its use in automobile catalytic converters. These devices control emissions by oxidizing or reducing toxic chemicals into less harmful substances. Most modern-day vehicles are fitted with a three-way converter. Oxidation works in two ways, converting carbon monoxide to carbon dioxide, and unburned hydrocarbons to carbon dioxide and water. Reduction converts harmful nitrogen oxides to nitrogen and water. Palladium is particularly efficient at removing unburned or partially burned hydrocarbons from fuel, especially during cold starts.

The element is also used in electronics in multilayer ceramic capacitors (electronic components that store electric charge), which are one of the most widely used forms of capacitor found in computers, cell phones, and televisions. Palladium is also a key element in modern jewelry; if you are wearing any white gold jewelry, it is likely to be an alloy of palladium and gold. The element has also been alloyed with silver and used in dentistry.

Palladium has an intriguing chemical trick up its sleeve—it has an unusual ability to store hydrogen gas. A solid chunk of the metal can absorb up to 900 times its volume in molecular hydrogen (H_2), without any external pressure. So how does this happen? When a hydrogen molecule meets the surface of the palladium, it breaks into its component atoms and is able to penetrate and pass through the palladium atoms. Hydrogen atoms can also diffuse right through palladium, to recombine as H_2, which makes this a useful way of separating and purifying hydrogen. It could also be an ideal way of storing large amounts of the gas, if palladium wasn't so pricey. There is, however, ongoing research attempting to find alloys that could achieve the same result at a much lower cost.

| | 46 |
| Pd | |

A sample of the rare silver-white metal
palladium. It is named not after the
venue for dodgy Saturday-night comedy
performances, but after the asteroid Pallas.

Silver

Category: transition metal
Atomic number: 47

Atomic weight: 107.8682
Color: brilliant white metallic
Phase: solid

Melting point: 1,764 °F (962 °C)
Boiling point: 3,924 °F (2,162 °C)
Crystal structure:
face-centered cubic

Silver invariably plays second fiddle to gold (see page 182), which is a shame as it possesses many intriguing qualities. Amazingly, silver is so ductile that 0.035 ounces (1 gram) of it can be stretched into a wire of more than 1.25 miles (2 km) in length. It is the best conductor of electricity of any element, and has the highest thermal conductivity of any metal. Soft, white, and lustrous, it is simply beautiful.

Like gold, silver is one of the least chemically reactive of the transition metals, although it does react with sulfur (see page 46), which is why it tarnishes. This has not detracted from its appeal over the centuries; slag heaps in Turkey and Greece indicate that silver was being mined as long ago as 3000 BC and it has been prized for use in ornaments, jewelry, tableware, and coinage for millennia. One of the most important statues remaining from ancient Greece features silver used in a subtle and beautiful form; look closely at the magnificent bronze statue of the Charioteer of Delphi (474 BC) and you'll notice that the proud racer has glinting silver teeth. Silver is mentioned in the Bible, often in relation to money; it features in both the Old Testament and the New Testament, which tells of how Judas Iscariot took the infamous bribe of 30 pieces of silver to betray Christ.

The name "silver" is derived from an Anglo-Saxon word, *siolfor*. Its chemical symbol, Ag, comes from the Latin name for silver: *argentum*. Silver was so important in Latin America that a country was named after it—Argentina. One of its rivers, the Rio

de la Plata, was also named after the Spanish word for silver, *plata*, because it was a rich alluvial source of the element; the 16th-century European invaders took little time in helping themselves to its treasure.

Pure silver is rarely found in its natural form. It's present in a number of naturally occurring minerals, including tetrahedrite and argentite, as well as lead, copper, and cobalt arsenide ores. Key mining areas are Mexico, Peru, the USA, Canada, and Australia. In 2011, silver mine output rose to a record high of 26,113 tons (23,689 tonnes).

In early times, silver was produced through a process called cupellation: ore or metal was heated in a cup while air was blasted over it; this would separate other metals, such as lead, iron, and copper, to leave a globule of pure molten silver. Today, silver is often produced as a by-product of electrorefining, where it is separated from other metals in an electrolytic cell (an electrolytic cell is formed by suspending both an electrically positive anode and a negative cathode in a chemical solution). Here, the impure metal is used as the anode and the refined metal is deposited on the cathode.

Silver is the most reflective of metals and this accounts for its use in mirrors, although many are now backed with aluminum (see page 38), which is much cheaper. Silver is also important for use in heat-reflecting products, such as glass for windows in homes and offices, and car windscreens.

Pure silver can be too soft for use in everyday objects, so copper is added to

Silver in pellet form. The element's applications include jewelry (as a precious metal), utensils, coins, electronic components, and mirrors. It has also proved to be a rather good investment, with the price in 2012 being over six times what it was in 2000.

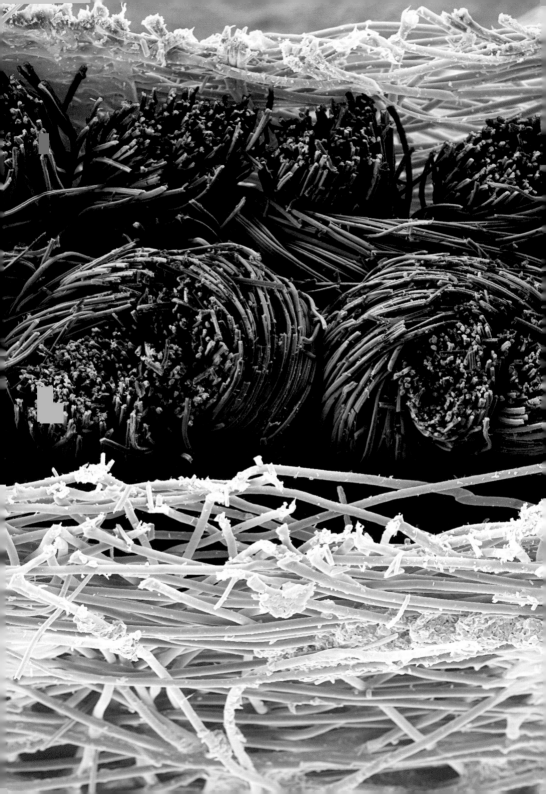

create sterling silver, which is both durable and decorative. Sterling silver trophies and tableware are usually 93 percent silver and seven percent copper; silver jewelry has a higher copper content, at 20 percent. Silver is also used in the production of high-quality musical wind instruments, and is favored for the mellow sound that it helps to create. Flutes for professional musicians are often made out of silver, with some commanding prices of between US$21,000 and US$55,000.

Silver tubes of a different kind have proved important in lifesaving surgery. For many years, silver pipes have been used in tracheostomies, a surgical procedure to aid breathing whereby an opening is created in the neck at the front of the trachea (windpipe) and a tube is inserted connected to an oxygen supply. Silver is used in tracheostomy tubes because it is not particularly reactive and is therefore less likely than many alternative materials to irritate the patient. However, plastic and other materials are increasingly being used because they are flexible and more comfortable.

A tracheostomy can help not only patients as an emergency procedure following an accident but also those with long-term illnesses, such as multiple sclerosis, that can make breathing difficult. The world's leading cosmologist, Professor Stephen Hawking, who suffers from a form of motor neurone disease, had this operation performed on him when his breathing became irregular in 1985. The operation saved his life.

Despite being relatively unreactive as metals go, silver will bond with some elements to form compounds with important medicinal uses. Silver ions and compounds have a toxic effect on some bacteria, viruses, and fungi, without being dangerous to humans. Physicians have been aware of the element's medicinal and germicidal properties since ancient times; the Greeks kept water and other liquids fresh in silver vessels, and the Romans stored wine in silver urns to prevent it from spoiling.

Silver will dissolve in nitric acid to form silver nitrate, a transparent photosensitive solid that is the basis of other compounds that have antiseptic properties and can be used to stain glass yellow. The compound was used in medieval times to treat warts and in the 19th century it was found to prevent blindness in babies following birth; administered as an eyedrop solution, it would kill the microbial infection that caused this common problem. Silver nitrate solution was used to prevent infection during the First World War, before antibiotics were introduced. Later, silver sulfadiazine topical cream became widely used for the antibacterial care of burns. Silver salts can also be added to swimming pools to sterilize them, instead of using chlorine.

As an interesting footnote ... silver can kill the sulfur-producing bacteria that cause smelly feet so, if you suffer from this problem, why not invest in some socks with silver thread woven into them?

A colored scanning electron micrograph (SEM) image of a sample of Actisorb Silver wound dressing. This consists of an activated charcoal cloth impregnated with silver. The charcoal and silver comprise the black layer, while the white layers are normal nylon fabric dressing. The silver has an antimicrobial action that kills bacteria in the wound. This dressing also absorbs toxins and reduces the wound's smell. Actisorb Silver is used on wounds including fungal lesions, infected pressure sores, and leg ulcers.

Cadmium

Category: transition metal
Atomic number: 48

Atomic weight: 112.411
Color: silvery blue
Phase: solid

Melting point: 610 °F (321 °C)
Boiling point: 1,413 °F (767 °C)
Crystal structure: hexagonal

Cadmium is so toxic that it features on the United Nations Environmental Program's list of top 10 hazardous pollutants. It is poisonous in an insidious manner and continued exposure to it causes a disease known in Japan as *itai-itai* (literally meaning "ouch-ouch"), for the joint and bone pain caused. The condition was first recognized in the Jinzu River Basin in the Toyama Prefecture, where land and water had become polluted by waste from a zinc mine. Rice grown in the area contained ten times more cadmium than normal rice. The disease was officially recognized in the late 1960s and locals successfully sued the mine companies for damages.

Cadmium claimed victims elsewhere during the 1960s. A group of British workers removing part of a construction tower on the Severn Road Bridge in 1966 used an oxyacetylene torch to melt the steel bolts that held it together. There was no ventilation in the tower and the job took all day. The following morning, all the workers felt ill and had difficulty in breathing. One man was hospitalized and died a week later. All his coworkers were also admitted to hospital for treatment but survived. The reason behind the tragedy was that the bolts they had removed were coated in cadmium, which had become activated by the heat.

Cadmium is a silvery metal with a bluish tinge and is soft enough to cut with a knife. It is in group 12 (making it a transition metal), along with mercury and zinc. The discovery of cadmium has its roots in zinc oxide production in the early 1800s; German chemists trying to obtain zinc oxide from calamine (also known as cadmia) noted that, when it was heated, sometimes a smelly yellow product was left instead of the pure white that they expected. In 1817, Inspector of Pharmacies Professor Friedrich Stromeyer of Göttingen University was asked to study the problem and he concluded that the yellow product must be a previously unknown element, which he separated and called cadmium after the ore.

Cadmium sulfide became used as a pigment and was commonly known as cadmium yellow, which incidentally became one of the French impressionist painter Monet's favored tones. The hue could be altered by adding sulfur and selenium to create brown, red, and orange. Due to the element's toxicity, however, cadmium sulfide is no longer added to artists' paints or inks.

One positive side to cadmium was its use in nickel-cadmium batteries, which could be recharged time and time again and were therefore considered beneficial to the environment. However, these have now been replaced by batteries that are lighter, more powerful, and less toxic.

Exposure to cadmium in general has come mostly from tobacco smoking. It crops up in unexpected places, however. In May 2006, fans of London's Arsenal soccer club were disappointed when a sale of the seats from the club's old stadium, Highbury, was cancelled; the seats were discovered to contain trace amounts of cadmium.

Cd 48

Cadmium is a soft, bluish-white transition metal. It is exceedingly poisonous. Despite this, it has found applications in rechargeable batteries, televisions, and bearing materials.

Indium

Category: post-transition metal
Atomic number: 49

Atomic weight: 114.818
Color: silvery gray
Phase: solid

Melting point: 314 °F (157 °C)
Boiling point: 3,762 °F (2,072 °C)
Crystal structure: tetragonal

Indium may not be a household name, but it has become an important component in the age of technology. This soft, lustrous, silvery metal is in group 13 of the periodic table and shares some properties with its upper and lower neighbors, gallium and thallium.

The element was discovered in 1863 by the German scientists Ferdinand Reich and Hieronymus Richter. Reich was studying the zinc ore sphalerite, which he believed might contain the recently discovered element thallium. Using a spectroscope, he examined a precipitate that he believed would yield thallium's distinctive green line. Reich was color blind, so he asked his colleague Richter to check the spectrum for him. The pair were in for a surprise. Instead of a green line, Richter saw an intensely beautiful violet–blue line that had not been noted before. Reich and Richter knew that they had found a new element and it was named after the Latin word *indicum*, meaning "violet" or "indigo."

Pure indium metal can be found in nature, but most of the element is created as a by-product of smelting zinc ore. The element didn't see much use until the Second World War, when it was used to coat bearings in high-performance aircraft. Today, however, it is in great demand in LCD televisions and computer monitors, in the form of indium tin oxide. This is a transparent conductor of electricity that allows signals be sent to individual pixels on the screen without blocking light from other pixels. Indium production has increased rapidly in the last decade, with China being the leading supplier.

Based on current consumption rates, scientists estimate that there may only be 13 years' supply of indium left. However, the Indium Corporation of America believes that the long-term supply is sustainable, by increasing the yields from current sources and broadening the range of base metals that indium can be extracted from.

Like gallium, indium can be used to wet glass and, if it is allowed to deposit, it will form a mirror as reflective as those made with silver and much less likely to corrode. Indium is also added to other metals to produce strong, hard alloys, some of which have been used in dentistry and to solder alloys to strengthen joints against thermal fatigue.

Indium is not known to be used by living organisms; however, in small doses its salts appear to boost metabolism. If you ingest more than just a few milligrams, though, the salts will produce a toxic reaction affecting the heart, liver, and kidneys. The radioactive isotope In-111 emits gamma radiation and is used in certain medical scans to differentiate between conditions such as osteomyelitis (an acute or chronic bone infection) and decubitus ulcers (pressure ulcers) to allow accurate diagnosis and treatment.

Here's a fascinating fact: did you know that if you bend a piece of indium metal, it lets out a high-pitched shriek? If you heard it, you wouldn't soon forget it. The "cry" is caused by the crystals inside the metal breaking and rearranging. The phenomenon also occurs when tin, indium's next-door neighbor in the periodic table, is bent.

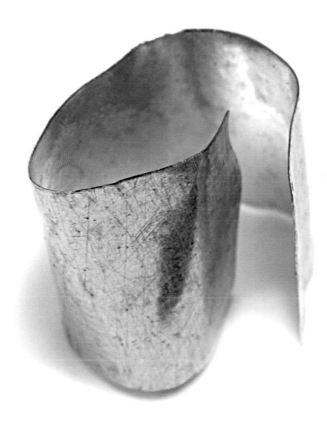

A strip of indium foil. The rare, soft malleable metal is used inside computers as an interface material between microchips and their heat sinks to dissipate thermal energy and prevent the chips from overheating.

Tin

Category: post-transition metal
Atomic number: 50

Atomic weight: 118.710
Color: silver–white
Phase: solid

Melting point: 449 °F (232 °C)
Boiling point: 4,716 °F (2,602 °C)
Crystal structure: tetragonal

Tin may seem a humble element, sometimes referred to as a "poor" metal because of its softness, but when it is alloyed with other elements it becomes very important indeed. Early man discovered that when tin was alloyed with copper, it produced bronze—a metal that was easy to work with (because it melted at a low temperature) and yet hard enough to make swords, arrowheads, and tools.

The Bronze Age began around 3300 BC and was an important stage in the development of civilization. Tin was mined in Turkey, Spain, France, and Britain, and traded around the Mediterranean Sea, as documented in the times of Julius Caesar. Bronze became highly prized and was used to create magnificent artworks, such as the Great Buddha of Kamakura in Japan and the Colossus at Rhodes—a 98-foot (30-meter)-high statue of Helios, the Greek sun god, which was destroyed during an earthquake in 226 BC. Pewter became another popular alloy of 85 to 90 percent tin with copper, antimony, and lead. It was used for flatware (tableware) from the Bronze Age until the 20th century.

Tin does not oxidize on exposure to air and as a result it is often used to coat other metals and prevent corrosion. The practice of tin-plating ironware is an ancient one and developed into a thriving industry in the Middle Ages, predominately in the Central European states of Bohemia and Saxony. By the 18th century, tin was used to plate cauldrons and bowls and to make items such as teapots, tankards, and plates.

In modern times, tin has been used to provide a corrosion-resistant plating for steel. Due to its low toxicity, tin-plated metal has been used extensively for food packaging, giving the name to tin cans, which are made mostly of steel. Tin has also been widely used as a solder, alloyed with lead, primarily for joining pipes or electric circuits. However, due to lead poisoning, use of this alloy has declined.

At very low temperatures, tin decays to gray dust and this is known as "tin pest" or "tin plague." The process occurs gradually at moderately low temperatures such as 14 °F (-10 °C), but accelerates when temperatures drop to -27 °F (-33 °C). Tin pest was commonly noted in the 18th century in Russia; in particularly harsh winters, the tin piping of church organs would become covered in scales that would crumble to the touch.

The deaths of Captain Robert Scott and his fellow explorers to the South Pole in 1912 have been attributed to tin pest. En route to the pole, Scott left stores of tinned food and canned paraffin for the long journey back. After reaching the pole—and discovering that Norwegian explorer Roald Amundsen's party had beaten them to it by five weeks—the exhausted men began their trek back across the ice, only to find that their stores of paraffin had drained away through tiny holes in the tin-soldered joints of the cans. Scott and his team eventually died of exposure.

Sn
50

Tin was one of the earliest metals used by humans. Today, one of its functions is as a principal ingredient of the alloy solder—used for joining pipes together in plumbing and for securing electronic components to circuit boards.

Antimony

Category: metalloid
Atomic number: 51

Atomic weight: 121.760
Color: silver–gray
Phase: solid

Melting point: 1,168 °F (631 °C)
Boiling point: 2,889 °F (1,587 °C)
Crystal structure: trigonal

Antimony quite literally means "not alone" when translated from its ancient Greek roots (*anti* and *monos*)—a reference to the fact that it is rarely found in its pure form in nature, instead existing as compounds with other elements. The name itself is thought to have first been used by Tunisian doctor Constantine the African in the 11th century. Its chemical symbol Sb comes from the Latin word *stibium*, a term for the only naturally occurring mineral containing antimony sulfide; the Swedish chemist Jöns Jakob Berzelius is believed to have been the first to abbreviate this to Sb, doing so during the 18th century.

Antimony's influence stretches back through history. The ancient Greek navy is believed to have used it in the recipe for Greek fire. The exact recipe is a mystery, but it is thought to have been a blend of crude oil, saltpeter (the oxidizing agent in gunpowder), and antimony sulfide (stibium)—which gives off large amounts of heat when it burns.

Antimony is toxic; ingesting large quantities of it produces symptoms similar to (but milder than) arsenic poisoning. Despite this, it's found many applications in medicine over the years. In the first century AD, the Romans are believed to have used stibium for burns and dermatological disorders. Medicinal use of antimony peaked in the 1700s. In particular, it was used as a nausea-inducing emetic. This was achieved by leaving a piece of antimony in a goblet of wine overnight (or leaving the wine in an antimony goblet). The wine would absorb enough of the element to do the job, but not so much as to be a toxic dose.

One theory ascribes the demise of the Austrian composer Mozart to an overdose of antimony tartrate, which the musical genius had been prescribed by his doctors but which he was reputed to take in doses far more liberal than intended. The symptoms of his death—swollen hands and feet, fever, and severe vomiting—exactly match those of antimony poisoning.

Indeed, once the true nature of antimony became clear, it soon migrated from the doctor's bag to that of the murderer. During the 19th century, many criminals went to the gallows having administered antimony to those to whom they had taken a dislike (in the hope their deaths would be attributed to mystery stomach complaints).

Recent times have seen new applications for the element. The discovery that antimony sulfide reflects infrared radiation in the same way as vegetation has led to its use in military camouflage paint. Antimony is unusual for a metal in that it expands upon solidifying (much like water) and this gave it an early use in printing press alloys. Tin-lead-antimony alloys expand to fill molds extremely well, giving excellently defined type. Antimony is also used in bearings, car batteries, flameproof plastics, and semiconductors.

The most bizarre application of antimony is the "reusable laxative." During the Middle Ages, pieces of antimony would be swallowed, the brief toxicity of which would stir up the bowels; and the laxative pellet could be recovered from the results for future use.

Sb 51

It's been suggested that this silvery metalloid element may have been responsible for the death of Mozart. As well as being a poison, it's found applications in areas as diverse as camouflage paint and semiconductors.

Tellurium

Category: metalloid
Atomic number: 52

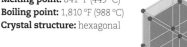

Atomic weight: 127.60
Color: silver–white
Phase: solid

Melting point: 841 °F (449 °C)
Boiling point: 1,810 °F (988 °C)
Crystal structure: hexagonal

Tellurium is in the same family as sulfur and selenium and shares some of their unpleasant, malodorous characteristics. Tellurium in its elemental form is not particularly toxic, but it can produce side effects such as bad breath and body odor. However, some compounds of tellurium can be far more dangerous. A small amount—0.07 ounces (2 g)—of sodium tellurite can kill, which was discovered by accident in 1946 when three soldiers were given a dose from a mislabeled bottle. The result was vomiting, internal bleeding, and respiratory failure. Long-term poisoning with smaller amounts results in "garlic breath" and tiredness.

The element was discovered at Sibiu in Romania, by Franz-Joseph Müller von Reichenstein in 1783. Müller collected rare minerals and became fascinated by one with an alluring metallic luster found in a mine at Zalatna. At first he thought it was antimony, but tests refuted this; he then inferred that it was bismuth sulfide. He was a little assumptive in announcing his hypothesis, as tests later showed that the mineral contained neither bismuth nor sulfur. Müller then concluded that it must be a gold ore with a previously undiscovered element.

Three years and many tests later, he was fairly sure that he had found a new element and asked chemists to verify this. They agreed and Müller published his findings in a Vienna journal. He also sent a sample to the leading German chemist Martin Klaproth (see also uranium, page 210) in 1796, who reiterated the other chemists' opinions. Müller soon obtained a pure sample of the element and called it tellurium (from Latin *tellus*, meaning "Earth").

Tellurium in its pure form is a brittle, silver-white metalloid, rarely found on Earth. Tellurium will burn in air and oxygen, but it is not affected by water. Minerals of it include calaverite (a compound of tellurium and gold), sylvanite (silver, gold, and tellurium), and tellurite (tellurium and oxygen), but none are mined; most tellurium is obtained from anode sludges left over from copper refining. It is also obtained as a gray powder from electrolytic reduction of sodium tellurite solution. The USA, Peru, Japan, and Canada are the main producers.

Tellurium makes a useful alloy with lead, to make it tougher, and with stainless steel and copper, to enhance their "machinability." It can also be used to vulcanize rubber. Tellurium suboxide is used in the media layer of rewritable DVD and Blu-ray disks, and also in memory chips, glass production, and solar cells.

The rising demand for tellurium resulted in a sharp increase in its value, from US$14 per pound (US$30 per kg) in 2,000 to US$150 per pound (US$330 per kg) in 2007. There is an ongoing debate over whether production of tellurium can keep pace with demand, particularly in the solar industry. Due to reduced copper production and new alternatives to copper extraction, availability of tellurium is decreasing globally, which will not help curtail the rocketing price of the element. Developments in thin film solar cell technology, using CdTe (cadmium telluride), could increase the demand for tellurium yet further in the future.

Te 52

Tellurium is a metalloid chemical element discovered in the late 18th century. It is used in the manufacture of DVDs, microchips, and glass.

Iodine

Category: halogen
Atomic number: 53

Atomic weight: 126.90447
Phase: solid
Color: black

Melting point: 237 °F (114 °C)
Boiling point: 364 °F (184 °C)
Crystal structure: orthorhombic

Iodine is essential to human beings and many animals, as a key constituent of thyroxine and tri-iodothyronine. These are produced by the thyroid gland in the neck and regulate a number of metabolic functions, including controlling our body temperature and the levels of other hormones. A deficiency in iodine results in diseases such as goiter, which is characterized by a swollen neck (due to enlargement of the thyroid gland), and hypothyroidism, when the invalid feels tired and cold, gains weight, and suffers from dry, itchy skin. A lack of iodine in pregnancy produces babies with a physical and mental condition known as cretinism.

If the body produces too much iodine, then hyperthyroidism results, with symptoms including restlessness, weight loss, heat intolerance, and difficulty in concentrating. Severe deficiency in iodine can lead to impairment of intellectual and physical development, which was a particular problem in developing countries until the late 1990s, when health agencies ran campaigns to have table salt iodized.

Iodine in its natural form is toxic but, like other members of the halogen group, it forms a safe and stable ion, iodide, which is highly soluble and in plentiful supply in seawater. The element is isolated by acidifying brine and bubbling it through chlorine gas.

It's important to note that often iodine is used as a blanket term when people are referring to its various forms. Elemental iodine is black, but most people associate the element with the violet tint that it produces

when it vaporizes. It's not surprising that its name is derived from this beautiful color, with its root in the Greek word *iodes*, meaning "violet."

Iodine was discovered during the Napoleonic Wars, when British forces blockaded France to cut off supplies of saltpeter (potassium nitrate) needed for gunpowder. Bernard Courtois set up a business using seaweed ash collected from the coasts of Normandy and Brittany, which he boiled to extract potassium chloride. Experimenting one day, he added sulfuric acid to the residual liquid and was astonished to see a cloud of purple vapor rise. Samples of the chemical were submitted to the Institut Impérial de France and it was agreed that a new element had been found.

Iodine has a radioactive isotope, I-131, which was released into the atmosphere by the explosion at the Chernobyl nuclear power plant in 1986. This resulted in a huge rise in thyroid cancer in children living in contaminated areas. I-131 has a half-life of just eight days, but this was long enough to pass into the food chain via grazing animals and milk. More than 4,000 people contracted thyroid cancer among the 18 million who were exposed to the fallout. Treatment for this form of cancer has a high success rate; intriguingly, part of the process involves giving a dose of radioactive iodine to kill the cancer cells.

Iodine is also an antiseptic disinfectant and is used to treat open wounds. Modern uses include iodine-based disinfectants, animal feed, printing inks, and dye.

I 53

Crystals of the chemical element iodine are heated
to produce violet-colored iodine vapor. Its name is
derived from the Greek for "violet." The element is
used in disinfectants and dyes.

Xenon

Category: noble gas
Atomic number: 54

Atomic weight: 131.29
Color: colorless
Phase: gas

Melting point: -169 °F (-112 °C)
Boiling point: -163 °F (-108 °C)
Crystal structure: n/a

Xenon is named after the Greek word *xenos*, meaning "stranger," which provides a heavy hint of how trying this element was to discover. It belongs in group 18 of the periodic table, with the other inert or noble gases (those that tend not to react chemically with other elements).

The group was discovered over a period of time during the late 19th century, starting with the discovery of helium, which was noted in spectral lines in 1868 by Jules Janssen in India, then Norman Lockyer and Edward Frankland in London. The British chemist Sir William Ramsay and his assistant William Travers isolated argon from liquid air in 1894, and went on to discover krypton and neon in 1898.

Later that year, thanks to the success of their experiments, Ramsay and Travers were given a liquid-air machine by the industrial chemist Ludwig Mond. Through repeated fractional distillation of liquid air, they were able to extract more krypton and eventually an even heavier gas: xenon. On examination in a vacuum tube, it produced a striking blue light and the team realized they had discovered another noble gas. They called it xenon, or stranger, because it had proved so hard to find.

Xenon was believed to be totally chemically inert until 1962, when British chemist Neil Bartlett and his team at the University of British Columbia, Canada, discovered it could form a compound with fluorine and platinum. More fluorine compounds followed and now xenon is known to also bond with hydrogen, sulfur, and gold—although these compounds are only stable at very low temperatures. Xenon has nine naturally occurring isotopes and only Xe-136 is radioactive.

Uses of xenon have been limited because it only occurs in trace amounts in Earth's atmosphere and is therefore expensive. Its commercial uses include gas-discharge lamps, more often known as "neon" lighting. It has also been used in the flash of modern cameras, especially in high-speed photography, due to its ability to provide an intense and almost instantaneous burst of light when pulsed with high-voltage electricity.

During the 1950s, xenon was considered for use as a general anesthetic because it produced fewer side effects than nitrous oxide, although this was not considered economically viable due to the element's expense. Thanks to advances in the recovery and recycling of xenon, the element may yet become viable for this use.

Xenon's applications extend into outer space, as it is an ideal form of fuel for the ion engines found in satellites and deep space probes. As the gas flows into the ion engine, its atoms are electrically charged to become ions. These can be moved around by electric voltage and accelerated to very high speed before they are shot out of the engine at around 24,000 miles (40,000 km) per second, creating powerful thrust. The speed can then be modified to keep the craft on course in outer space. Xenon Ion Propulsion Systems (XIPS) are 10 times more efficient than conventional propulsion units.

Xenon is a colorless, odorless, inert, noble gas. Hard to believe, then, that it's used as rocket fuel. A relatively new form of rocket called an "ion engine" uses an electric field to accelerate charged xenon atoms to form a high-speed exhaust jet that can propel a vehicle through space.

Cesium

Category: alkali metal
Atomic number: 55

Atomic weight: 132.90545
Color: silvery gold
Phase: solid

Melting point: 83 °F (28 °C)
Boiling point: 1,240 °F (671 °C)
Crystal structure: body-centered cubic

What fun chemists can have in the laboratory with the alkali metal group. The group becomes progressively more reactive as you look down its column on the periodic table and cesium is no disappointment. In its pure form, it will explode violently when dropped into water and it is so reactive that it has to be stored in oil and handled in an inert atmosphere.

This soft, silvery-gold metal has a low melting point and becomes liquid at just above room temperature. Pollucite is the main mineral and this is quite rare. Production of cesium amounts to around 22 tons (20 tonnes) per year. Most is sourced from Bernic Lake, Manitoba, Canada, and some from Southwest Africa and Zimbabwe.

Cesium was discovered in 1860 by the German scientists Robert Bunsen and Gustav Kirchhoff, who found it in a sample of mineral water from Durkheim. They boiled the water to remove other salts and sprayed the remaining liquid into a flame, so that the light could be analysed by a spectroscope. This showed two blue lines that had never been seen before, so Bunsen and Kirchhoff were convinced that they had discovered a new element. They were later able to extract the pure metal, which was named cesium after the Latin word *caesius,* meaning "sky blue." The following year, the duo also used spectroscopy to discover rubidium.

Like rubidium, cesium is used in atomic clocks, which are the most accurate time devices known to man. The cesium atomic clock depends on the electromagnetic transitions between two states of Cs-133 atoms and uses this to define a very accurate reference time interval. The first cesium clock was built in 1955 by Louis Essen at the National Physical Laboratory in the United Kingdom, and the device has been improved since then to measure time to an accuracy of one second in 300,000 years. A number of atomic clocks send signals all over the world to standardize time for cell phone networks, the internet, and satellite navigation systems.

Cesium has two radioactive isotopes, Cs-134 and Cs-137, which became notorious following their release into the atmosphere during above-ground testing of nuclear weapons from 1945 to 1963. The explosion at Chernobyl nuclear power plant in the Ukraine in 1986 also released a large amount of Cs-137, which has a half-life of 30 years. The fallout from the disaster spread across Western Europe as far as the Republic of Ireland and affected large numbers of livestock after it was washed into the soil and absorbed by plants. Government restrictions were imposed on areas affected by the fallout and livestock could not be sold for slaughter until they had been checked for contamination. Restrictions were still in place in some areas of Wales 25 years after the explosion.

Japan's Fukushima Daiichi nuclear plant, which was damaged following a tsunami in 2011, released Cs-137 at a level approaching that of Chernobyl. Researchers have commented that fallout from the Japanese plant poses a lesser risk than Chernobyl because the radioactive particles bubbled off the damaged fuel instead of being blasted out in smoke that drifted across a whole continent.

A vial containing the alkali metal cesium. Like sodium, potassium, and the other elements in Group 1 of the periodic table, cesium reacts violently when dropped into a bowl of water. In fact, cesium will blow the bowl to bits.

Barium

Category: alkaline earth metal
Atomic number: 56

Atomic weight: 137.327
Color: silver–gray
Phase: solid

Melting point: 1,344 °F (729 °C)
Boiling point: 2,979 °F (1,637 °C)
Crystal structure: body-centered cubic

Barium is a soft, silvery alkaline earth metal named from the Greek word *barys*, meaning "heavy," and while barium may not be heavy in pure form, many of its minerals and compounds are very dense. Barium sulfate is so dense that it is used in oil-well drilling, where a slurry of the compound is pumped into the bore hole and the pressure created forces rock chips up and out of the hole.

Barium is highly reactive with air and does not occur naturally in its pure form, but its compounds are plentiful on Earth. In the early 1600s, barium, in the form of the mineral barite, attracted attention in Bologna, Italy. Alchemist Vincenzo Casciarolo found some shiny pebbles, heated them up, and found that they glowed in the dark. If the stones were exposed to strong sunshine during the day, they would glow the following night. "Bologna stone" attracted widespread attention and in the 18th century it was investigated by chemists, who realized that it contained the sulfate of a hitherto unknown element.

In the 1770s, barium was detected in the ash of plants and vegetables and in the 19th century, it was found in seaweed and some sea life. Elemental barium was first isolated by Sir Humphry Davy in 1808, who separated molten barium hydroxide using electrolysis.

The main barium ores are barite (barium sulfate) and witherite (barium carbonate), both of which are mined mostly in the UK, USA, Germany, Italy, and the Czech Republic. Due to its highly reactive nature, barium is used as a "getter" to scavenge the final traces of oxygen and other gases and so form a perfect vacuum. Soluble barium salts are poisonous and if they're ingested lead to vomiting, diarrhea, paralysis, and cardiac irregularities. In 1993, Texan teenager Marie Robards poisoned her father by stealing barium acetate from her school chemistry laboratory and mixing it with his dinner. Robards' parents were divorced and she stated that she killed her father "so that she could go and live with her mother."

The more commonly known—and far less dangerous—"barium meal" has been used extensively in medicine to diagnose gastric and intestinal disorders. As an atom, barium has an electron configuration that readily absorbs X-rays, meaning that it shows up clearly on medical scans. The compound barium sulfate is used for this procedure because it is insoluble and will not react with stomach acid. The barium meal is swallowed an hour prior to the X-ray and shows the form of the stomach and intestines. Use of barium meals is declining due to the increasing practice of endoscopy to examine the esophagus and stomach.

Most barium compounds find niche uses mainly due to their alarming toxicity. Barium carbonate is insoluble in water but it dissolves in stomach acid and has been used as rat poison. This compound is also used in glassmaking, where it increases the luster of the glass. Visit a firework display and the brilliant green pyrotechnics that you see may also be the product of barium compounds.

A chunk of barite, a mineral made from the compound barium sulfate. Barium is a good absorber of X-rays, which is why this element is probably best known for the "barium meals" used in medicine to diagnose disorders of the digestive tract.

Lanthanum

Category: lanthanide
Atomic number: 57

Atomic weight: 138.90547
Color: silver–white
Phase: solid

Melting point: 1,688 °F (920 °C)
Boiling point: 6,267 °F (3,464 °C)
Crystal structure: hexagonal

Lanthanum is the first in a sequence of 15 chemical elements known collectively as the lanthanides. These traditionally sit beneath the main periodic table in one of two strips (the other, directly below it, being the actinide sequence). The sequence stretches from the element lanthanum (atomic number 57) up to lutetium (atomic number 71)—strictly speaking lutetium isn't a lanthanide, but it is usually included because of its chemical similarity to the "proper" lanthanide elements. The lanthanides are appended at the foot of the table merely for convenience—to prevent the table from being a letterbox-format 15 elements wider than it currently is.

The lanthanides are sometimes known as the "rare earth metals," although this is now an obsolete term. When they were first discovered, these elements were indeed rare compared to the other earth metals of the time. Now, however, they are relatively abundant compared to the other known elements. Lanthanum, for example, is three times more common than lead, and as common as lead and tin combined, being present in the Earth's crust at a concentration of 32 parts per million.

The lanthanides are grouped into a family of elements because their chemical properties are so similar. This is unusual for a collection of elements arranged horizontally on the periodic table—normally chemical similarities manifest in elements occupying the same vertical group. In the case of the lanthanides, the similarity arises because all the lanthanide elements have the same number of electrons in their outer orbit;

it's these outer electrons that typically interact with other atoms and are thus responsible for an element's chemical properties. Each lanthanide element has a different number of electrons in total but, as atomic number increases, these extra electrons are added to inner orbits. As a result, the same three outer electrons interact with the rest of the world. This is what makes the chemical properties of all the lanthanide elements much the same.

Lanthanum was discovered in 1839 by the Swedish chemist Carl Gustav Mosander; given its similarity to all the other elements in the sequence, it was no mean feat for the time. He heated cerium nitrate—cerium had been discovered in 1803—to make cerium oxide. He then added nitric acid and boiled off the liquid to leave the nitrate of a new element, which, on the advice of a friend, he named lanthanum from the Greek *lanthanein*, meaning "hidden."

At the time, the name may have referred to the rareness of these "rare earth" elements; now, it better represents the similarity of the elements, which makes them so difficult to separate from one another when brought up from the ground in their parent minerals. There are no minerals that contain just lanthanum, but a set of minerals that contain almost all the lanthanides. Unsurprisingly, this has led to some confusion over claimed discoveries of new lanthanide elements—for example, where a mixture of two elements has been mistaken for a single new one.

In its purified state, lanthanum is a soft silver–white metal that swiftly oxidizes

A sample of the rare earth metal lanthanum. The element is used in cigarette lighter flints and hybrid car batteries, and played a small part in the discovery of nuclear fission.

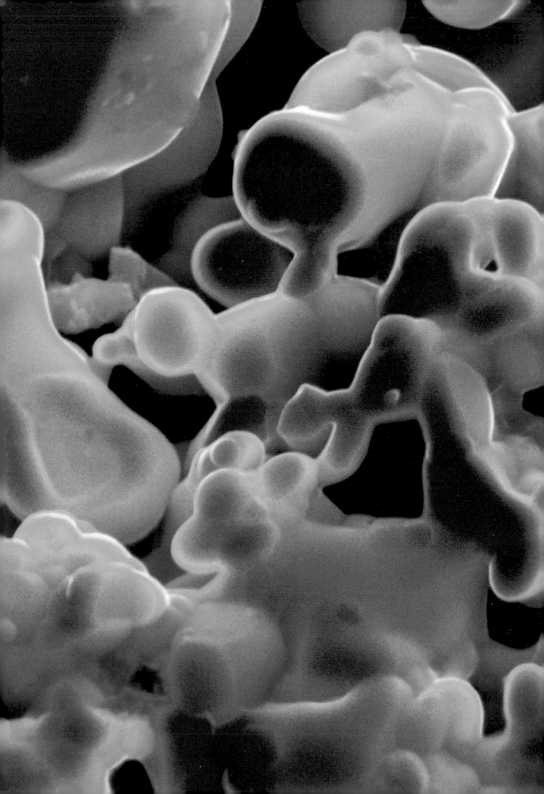

upon contact with the air. The metal itself wasn't isolated until 1923. Lanthanum is never found naturally in such a pure state, instead occurring within two minerals—monazite and bastnasite. These minerals contain nearly all of the lathanides, but up to 38 percent of their lanthanide content is in the form of lanthanum. Bastnasite is the richer mineral, but for many years monazite was much more common—that is, until the discovery of large amounts of bastnasite at Mountain Pass Mine in California, USA, in 1949. Other deposits have been discovered since in Africa and China.

World reserves of lanthanide elements are believed to total some 110 million tons (110 million tonnes), with around 77,000 tons (70,000 tonnes) being extracted from the ground each year. Due to the complexities of separation—owing in turn to the very similar properties of the elements—most is used in the form of a lanthanide alloy (mischmetal), in whatever proportions it happened to come out of the ground. Typically this is 50 percent cerium, 25 percent lanthanum, and 15 percent neodymium, with the other lanthanide elements together comprising around 10 percent. Generally speaking, heavier lanthanides are rarer because they have sunk down into the Earth's mantle, making them less common in the crust.

Lanthanum has many commercial applications: mischmetal plus iron is used to make cigarette lighter "flints"; and hybrid cars, such as the Toyota Prius™, use many kilograms of lanthanum in their nickel–hydride batteries. There may be further applications in eco-friendly power generation—a lanthanum–nickel alloy has the curious property of being able to soak up large quantities of hydrogen gas (up to 400 times its own volume). Hydrogen is one candidate future fuel that could be used to store and release electricity. Unlike a battery, which takes hours to charge, a car could be fueled with hydrogen in the same time it takes to fill a tank of gas—with the only by-product being water, released at the same time as the electricity.

Lanthanum is also effective at neutralizing phosphorous in a biological context. For this reason, it is used in ponds, where phosphorous fuels the growth of unwanted algae. Meanwhile, lanthanum carbonate is a treatment for kidney problems caused by excess phosphorous in the blood. Lanthanum can also be added to glass to improve its optical properties, boosting its refractive index and making it more resistant to corrosion, and to steel, making it more durable and deformable.

There have also been a few one-off applications. For instance, it was the discovery of unexpected lanthanides within a sample of uranium that, in 1938, led Otto Hahn and Lise Meitner to the discovery of nuclear fission—the uranium atoms were fissioning (breaking apart) to form atoms of other elements, some of which were lanthanides.

There are two main naturally occurring isotopes of lanthanum—with 138 and 139 particles in the nucleus. La-139 is stable, whereas La-138 is radioactive, with a half-life of 100 billion years. There are another 38 artificial isotopes that can be made in the lab—most of these have half lives of less than a day.

Lanthanum has been associated with mild toxicity in animal tests, leading to elevated blood sugar and low blood pressure and impacting on the function of the spleen and liver. Some people exposed to lanthanum compounds emitted from carbon arc lights have subsequently developed the debilitating lung disease pneumoconiosis.

This colored scanning electron micrograph (SEM) shows a ceramic, high-temperature superconductor. This superconducting metal material is made up of lanthanum, barium, and copper oxide. Superconductors carry electrical current with little or no resistance. The phenomenon occurs in certain materials at very low temperatures (just a few degrees above absolute zero), where the atoms in the material undergo a change allowing the electrons that carry electric current to slip past unimpeded.

Cerium

Category: lanthanide
Atomic number: 58

Atomic weight: 140.116
Color: iron gray
Phase: solid

Melting point: 1,463 °F (795 °C)
Boiling point: 6,229 °F (3,443 °C)
Crystal structure: face-centered cubic

Cerium is the second element in the lanthanide sequence—a block of 15 elements usually displayed as a horizontal strip at the foot of the periodic table, of which lanthanum (see page 134) is the first. Swedish chemists Jöns Jakob Berzelius and Wilhelm Hisinger discovered cerium in 1803. It was named after the asteroid Ceres (technically, a dwarf planet in the current classification scheme), which had itself only been discovered two years earlier.

Because of their similarity to one another, most of the lanthanide elements were discovered within the same mineral ores. However, cerium was different. It was found in so-called cerium salts, minerals made almost entirely of cerium. One of them, cerium silicate—or cerite—was originally believed to be an ore of tungsten when that metal was first discovered in the mid-18th century. However, Berzelius and Hisinger's analysis later revealed it to be an ore of a new element. Cerium in its pure metallic form wasn't extracted until 1875, using electrochemical techniques.

Cerium is the most abundant of all the lanthanides. Occurring with a concentration in the Earth's crust of 46 parts per million (by mass), it is almost as common as copper. The element is commercially extracted from the same ores as other lanthanides—such as bastnasite and monazite—in a process that involves crushing, heating, and treating with acid.

Cerium is something of a green element—not in terms of its actual color, but in the sense that it offers a range of applications that are beneficial to the environment.

Cerium sulfide has a deep red color that makes it excellent as a red pigment, replacing other more toxic chemicals such as cadmium, mercury, and lead. Cerium oxide is used in the manufacture of catalytic converters, forming part of the catalyst in car engine exhausts that stores and releases the oxygen needed to convert harmful carbon monoxide to carbon dioxide. Meanwhile, in diesel engines, adding a small amount of cerium to the fuel greatly lowers the quantity of particulates (solid pollutant particles) released into the atmosphere by encouraging these particles to burn in the engine.

There are a host of other applications, too. Cerium is used to coat the walls of self-cleaning ovens, where it helps oxidize cooking residues, converting them into ashlike compounds that are easily wiped away. It was used in cathode-ray TV tubes to reduce the screen darkening caused by continuous bombardment from high-energy electrons, and a form of cerium oxide was used to make the first gas mantles. Running a knife along a piece of cerium produces an impressive shower of sparks, which can be used to kindle fires (a property known as "pyrophorism")—leading to applications in lighter flints as well as campfire starters. Administered in small quantities orally, cerium even has medicinal properties as an antinausea drug.

There are four stable isotopes of cerium (with atomic masses 136, 138, 140, and 142). Of these, Ce-140 makes up 88.5 percent of natural deposits. There are 26 radioactive isotopes. Cerium exhibits moderate toxicity, causing skin irritation in humans.

A sample of the most abundant lanthanide element, cerium. Splinters of cerium are "pyrophoric" (igniting spontaneously on contact with air) which is why this element is used in campfire starter rods.

Praseodymium

Category: lanthanide
Atomic number: 59

Atomic weight: 140.90765
Color: silver–gray
Phase: solid

Melting point: 1,715 °F (935 °C)
Boiling point: 6,368 °F (3,520 °C)
Crystal structure: hexagonal

This somewhat verbosely named metal was discovered as part of another element, which turned out not to be an element at all. In 1841, the Swedish chemist Carl Gustav Mosander extracted the element lanthanum from a cerium-based compound. However, that wasn't all. Mosander believed that the residue left behind after removal of the lanthanum was itself another element, which he named dydimium. In 1882, spectral analysis led to suspicions that dydimium was not a pure element. Just a few years after that, in 1885, the Austrian chemist Carl Auer von Welsbach confirmed those suspicions when he was able to split dydimium into two new elements: neodymium and praseodymium.

The name *praseodymium* derives from Greek, meaning "green twin"—green because the element oxidizes to form a green layer on its exterior, pieces of which readily flake away. Praseodymium has atomic number 59, placing it within the lanthanide sequence of elements (also known as the "rare earths"), which lie in a horizontal strip near the bottom of the periodic table. Physically, the element is a silver–gray metal that's both malleable and ductile (meaning that it is easily squashed flat and stretched into wires, respectively).

Despite being a "rare earth" element, praseodymium is quite common, occurring with a natural concentration of 9.5 parts per million in the Earth's crust—making it four times more common than tin. It has just one naturally occurring isotope, Pr-141. This isotope is stable, though many radioactive isotopes, with mass numbers ranging between 121 and 159, are produced in nuclear reactions. The most stable, Pr-143, has a half-life of 13.5 days—most are under 10 minutes. Annual global production of praseodymium oxide is 2,755 tons (2,500 tonnes), with only a small proportion of this being converted into pure metal. The element is extracted from the minerals monazite and bastnasite, accounting for about five percent of the lanthanides found within them.

One of praseodymium's most important applications is in high-durability components for aircraft engines, where the element can be found alloyed with magnesium to form a superstrong construction material. Another important use is in welders' masks. Here, a small quantity of praseodymium added to glass filters out infrared heat radiation, and efficiently blocks the bright yellow light from flames that would force us to look away. Other uses include the cores of carbon arc lights for the movie industry and a coloring for cubic zirconia imitation gemstones to resemble the green jewel peridot.

Praseodymium has even been used to create the coldest temperature on Earth; the element is a so-called magnetocaloric material, meaning that its temperature is lowered by a magnetic field. This has allowed scientists to cool it to within one-thousandth of a degree Celsius of absolute zero: -273.15 °C (-460 °F). Meanwhile, in fundamental science research, praseodymium has been added to glass to slow the speed of light within it from 186,000 miles (300,000 km) per second to just a few hundred.

A piece of the dark "rare earth" metal praseodymium—although, far from being rare, this element is four times more common in the Earth's crust than tin.

Neodymium

Category: lanthanide
Atomic number: 60

Atomic weight: 144.242
Color: silver–white
Phase: solid

Melting point: 1,875 °F (1,024 °C)
Boiling point: 5,565 °F (3,074 °C)
Crystal structure: hexagonal

Neodymium was one of two chemical elements to be extracted from a substance known as "dydimium." When it was discovered in the 1840s, dydimium itself was initially thought to be an element, but studies soon alerted chemists to the fact that it wasn't a fundamental substance at all. In 1885, the Austrian chemist Carl Auer von Welsbach confirmed this when he split dydimium into two (genuinely) new elements, which were named praseodymium and neodymium.

The name *neodymium* means "new twin" in Greek. The element is part of the lanthanide series—a strip of elements normally displayed at the bottom of the periodic table for the sake of neatness. It is the second most abundant of the lanthanides and almost as common as copper (despite the lanthanides' alternative historical—and now largely defunct—name: the "rare earth" elements). Like the other lanthanides, it is predominantly extracted from the minerals bastnasite and monazite—of which it accounts for between 10 and 18 percent. These minerals are mined from sites in Australia, China, Brazil, India, Sri Lanka, and the USA.

The element has five stable isotopes, with atomic masses 142, 143, 145, 146, and 148. There are also two naturally occurring radioisotopes, Nd-144 and Nd-150, with half-lives of 2.29 million billion years and 7 billion billion years—roughly 1.5 million times and 5 billion times the present age of the Universe, respectively.

Neodymium is important in the manufacture of extremely high-power magnets. In 1983, it was discovered that an alloy of neodymium, iron, and boron (known as NIB) has magnetic properties far outstripping any other known permanent magnets. So much so that NIB magnets are used in a multitude of applications requiring small yet powerful magnetic sources—such as in-ear headphones, computer hard drives, and even in model aircraft engines, where electric motors incorporating NIB magnets are now replacing internal combustion engines. Using neodymium in the magnets powering electric motors has led to their use in hybrid cars, where the name of the game is building an electric car engine to compete with those that run on gas. The electric motors used to drive the Toyota Prius hybrid car each require 2.2 pounds (1 kg) of neodymium in their magnets.

There is one recorded case of two NIB magnets separated by 1.5 feet (0.5 m) snapping together with such force that they severed the tip of an intervening finger. If swallowed, the force pulling two of these magnets together from different sections of the digestive tract could lead to serious injury.

In addition to its magnetic properties, neodymium is also used as an additive in glass (to give the glass an attractive blue-purple hue), for coloring the light from artificial bulbs, and for making glass that can block harmful types of light (such as infrared—"heat" radiation). Meanwhile, the element has also found a use in lasers—so-called Nd-YAG lasers employ an alloy of yttrium-aluminum garnet and neodymium to generate a beam of intense coherent light, which in turn has myriad applications in industry and consumer technology.

Nd 60

Neodymium is a soft silvery, metallic element. It is used in magnets—alloyed with other metals it can make a magnet that is more powerful than those produced from any other substance.

Promethium

Category: lanthanide
Atomic number: 61

Atomic weight: 145
Color: silver
Phase: solid

Melting point: 1,908 °F (1,042 °C)
Boiling point: 5,432 °F (3,000 °C)
Crystal structure: hexagonal

At the foot of the periodic table is a strip of elements that's been moved from its proper location between atomic numbers 57 and 71 in order to keep the table neat. This group of elements is known as the lanthanide sequence, after its first member: element 57, lanthanum. The sequence, however, has an alternative name: the rare earth elements. Generally speaking, this is a misnomer; the lanthanides are usually anything but rare. One exception to that rule, however, is the element promethium.

Naturally occurring promethium is practically nonexistent on Earth, for the simple reason that the element is highly radioactive. Most elements with atomic numbers less than 83 (bismuth, page 192) are stable, or nonradioactive. The only two exceptions are technetium (page 104) and promethium. This is due to a quirk in the way protons and neutrons arrange themselves in the nuclei of these two elements, making it impossible for the particles to shuffle down into a stable configuration.

The most stable radioactive isotope of promethium (Pm-145) has a half-life of 17.7 years; however, for most of the isotopes this figure is less than 10 minutes. That means that radioactive decay causes the amount of Pm-145 in a sample to halve every 17.7 years and so a million tons of even the most stable promethium isotope will decay to just 0.04 ounces (1 g) in a little over 700 years. The Earth is an estimated 4.5 billion years old, and so any promethium that it was born with, will have long since decayed to virtually nothing.

In 1914, British physicist Henry Moseley realized that there was a gap in the periodic table between neodymium (atomic number 60) and samarium (atomic number 62). In the 1920s, two groups of researchers claimed to have found the missing element 61. In 1926, a team in Florence, Italy, claimed to have isolated it from the mineral monazite. The same year, a similar claim was made by scientists at the University of Illinois. The proposed elements were even named "florentium" and "illinium."

However, element 61 wasn't truly discovered until 1945, when scientists at Oak Ridge National Laboratory in Tennessee were sifting through the debris from one of the world's early nuclear fission reactors; promethium is one of the decay products of uranium. A substantial quantity of the pure metal wasn't obtained until later still; it took until 1963 for 0.35 ounces (10 g) of the element to be synthesized.

The 1945 team that was responsible for promethium's discovery included researcher Charles Coryell. It was his wife, Grace, who came up with the name "promethium," after Prometheus—the Titan of Greek legend who stole fire from the gods of Mount Olympus and gave it to mankind.

Promethium was, for a time, used for the luminous dials of watches—though its short half-life made for only a brief lifespan. Other uses are in batteries—for example, in spacecraft—and for predictable radiation emitters.

Radioactive promethium is an active ingredient in the manufacture of luminescent paint.

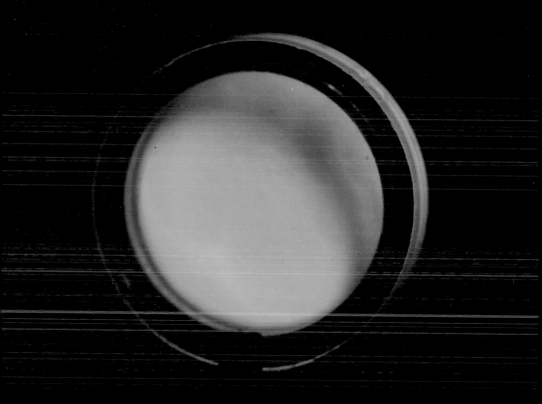

Samarium

Category: lanthanide
Atomic number: 62

Atomic weight: 150.36
Color: silver–white
Phase: solid

Melting point: 1,962 °F (1,972 °C)
Boiling point: 3,261 °F (1,794 °C)
Crystal structure: rhombohedral

Samarium is widely regarded as the first chemical element to be named after a person —albeit indirectly. The element is actually named after the mineral samarskite, but this in turn was so-called after the Russian engineer Vasili Samarsky-Bykhovets.

Samarsky-Bykhovets didn't personally have a hand in the discovery of samarskite, but in his capacity as head of the Russian Mining Engineering Corps, he allowed access to mineral samples for the German mineralogist, Gustav Rose. In 1839, Rose found a new mineral, which in gratitude he named after Samarsky-Bykhovets. In 1879, the French chemist Paul Émile Lecoq de Boisbaudran (see also gallium, page 78) succeeded in extracting the mineral dydimium from samarskite. Dydimium was once believed to be an element in its own right but later proved to be a blend of the elements praseodymium and neodymium. The same year, Boisbaudran was able to extract another chemical from samarskite, which detailed tests soon revealed to be a genuine new element. It was named samarium.

Samarium is a member of the lanthanide group—a sequence of 15 chemical elements that occupies a strip at the bottom of the periodic table; it oxidizes in moist air, forming a crust on its exterior, and bursts into flame at a temperature of 302 °F (150 °C).

Despite its initial discovery in samarskite, the mineral monazite is the richest known source today, with samarium accounting for three percent of the mineral's mass. Mining sites exist in Australia, Sri Lanka, and China.

Samarium was initially used in magnets, enabling some of the most powerful magnets ever developed until the development of neodymium-based magnetic alloys. Even so, samarium-cobalt magnets remain 10,000 times more powerful than those based on iron. Unlike neodymium alloys, magnets based on samarium don't lose their magnetism at high temperature— remaining magnetic above 1,292 °F (700 °C). Samarium-cobalt magnets are still used in the pickups for electric guitars.

Samarium has some unusual electrical properties. The compound samarium sulfide comes in the form of black crystals, which exhibit semiconducting properties—their behavior is somewhere between that of a conductor and an insulating material. However, under the application of 20–30 times atmospheric pressure it converts into a metallic solid, conducting electricity readily, and assuming a lustrous golden hue.

There are seven naturally occurring samarium isotopes, three of them radioactive: Sm-147, 148, and 149 all have half-lives of hundreds of billions of years. However, there are also manufactured isotopes; of these, Sm-153 is probably the most important—its radiation makes it an effective killer of cancer cells.

Samarium has proven useful not just for its emission of radiation but also for its ability to absorb it. Sm-149 readily soaks up neutrons— the particles triggering nuclear fission reactions. It is therefore used as a moderator in the control rods that govern the rate of energy production inside a nuclear reactor.

Like its sibling element neodymium, samarium
has desirable magnetic properties. It also has
an appetite for soaking up neutrons, leading to
its use in nuclear reactor control rods.

Europium

Category: lanthanide
Atomic number: 63

Atomic weight: 151.964
Color: silver–white
Phase: solid

Melting point: 1,519 °F (826 °C)
Boiling point: 2,784 °F (1,529 °C)
Crystal structure: body-centered cubic

Anyone who enjoyed watching color television before the rise of flat-panel displays should thank the element europium; the chemical was essential in the development of effective red phosphors, used for coating the insides of cathode-ray tube screens. Each point on a cathode-ray tube screen is a cluster of red, green, and blue phosphor dots. Bombarding the phosphors with a beam of electrons causes them to glow—a phenomenon known as fluorescence. The color at each point is then adjusted, using an electric field that aims the beam at the correct-colored phosphor or combination of phosphors.

Making the luminescent phosphors for green and blue colors was easy, using zinc sulfide with the addition of copper and silver, respectively. But red phosphors were initially so poor that the brightness of the other colors had to be diminished in order to maintain the correct color balance. That all changed with the discovery of europium-based phosphors, which emit copiously in the red portion of the visual light spectrum. The small number of cathode-ray tubes still produced today use red phosphors made from yttrium oxide-sulfide doped with a small quantity of europium.

The element's ability to give off red light has led to it also being used in compact fluorescent (energy-saving) lightbulbs, where the addition of some red color softens the usual harsh white of fluorescent strip lighting. Europium has further applications in lasers, and in superconductors—materials in which electrical resistance completely vanishes.

Europium is one of the lanthanide, or "rare earth," elements, seen at the foot of the periodic table. It was discovered in 1901 by the French chemist Eugène-Anatole Demarcay. He deduced that samples of the newly found element samarium were contaminated with small quantities of another element. Through a protracted sequence of chemical reactions, he was able to isolate what became known as europium, named after the continent of Europe.

Europium is a soft metal, silver in color. It burns at 356 °F (180 °C), oxidizes swiftly on contact with the air, and reacts readily with water. It's one of the rarer lanthanide elements, with just 110 tons (100 tonnes) of the pure metal produced annually, using minerals extracted from mines in China and the USA. Sources of europium include the rocks bastnasite and monazite, as well as the dark gray mineral loparite and the yellow-brown xenotime.

Naturally occurring europium exists as one of two stable isotopes, with atomic masses of 151 and 153. Europium is something of an oddity, in that its concentration in minerals, relative to the other lanthanide elements, is often anomalously large or small. This "europium anomaly" has been especially evident in samples of rock brought back from the Moon. Rocks gathered from the white lunar "highlands" show a marked positive anomaly (excess europium), whereas those recovered from the darker "maria" present an anomaly that's strongly negative. Some commentators have interpreted this as evidence that the Earth and Moon did not form from the same source of material.

Before we all got flat-screen TVs, color televisions
were bulky affairs made with heavy glass screens.
Essential in their construction was the element
europium, which was used to make the red
component of the color picture.

Gadolinium

Category: lanthanide
Atomic number: 64

Atomic weight: 157.25
Color: silver
Phase: solid

Melting point: 2,394 °F (1,312 °C)
Boiling point: 5,923 °F (3,273 °C)
Crystal structure: hexagonal

Gadolinium is one of the more common elements in the lanthanide series. It was discovered in 1880 by the Swiss chemist, Jean Charles Galissard de Marignac, who found a new oxide lurking in the mineral didymium. This mineral was itself once believed to be a pure element but was later proved to be a mixture made predominantly of two other elements (those came to be known as praseodymium and neodymium). In 1880, Marignac was able to isolate the oxide of another, genuinely new, element from didymium. In 1886, French chemist Paul Émile Lecoq de Boisbaudran was able to extract the pure element from the oxide. It was Boisbaudran who suggested the name "gadolinium," after the mineral gadolinite—a plentiful source of this element—and Marignac agreed. Gadolinite was first discovered near the Swedish village of Ytterby in 1780 and is itself named after the Finnish chemist Johan Gadolin, who was the first to describe it.

Approximately 440 tons (400 tonnes) of gadolinium metal are extracted from the ground every year, from reserves believed to total 1 million tons. It is excavated from mines in Australia, China, Brazil, and Sri Lanka—to name just a few. The pure element and its oxide are normally extracted from the raw minerals bastnasite and monazite, through a complex chemical procedure. Gadolinium has six naturally occurring stable isotopes (with atomic numbers 154–158 and 160) and one that is radioactive (Gd-152, which has a half-life of 1.1 trillion years, many thousand times the age of the universe). There are also a plethora of manufactured radioactive isotopes of gadolinium, with half-lives ranging from millions of years to just a few tens of seconds.

Gadolinium is the best neutron absorber of any stable element. Only one known elemental isotope has a higher absorbency: the unstable xenon-135. This has led to gadolinium's use as a neutron absorber in nuclear fission reactors (where neutrons are the particles responsible for perpetuating the chain reactions) and in medical imaging, where a gadolinium screen bombarded with neutrons briefly gives off penetrating radiation that can be used to create internal images of patients.

Gadolinium has interesting magnetic properties. It is ferromagnetic (meaning that it's attracted to a magnet)—until, that is, its temperature is raised above 66 °F (19 °C), at which point the magnetic properties promptly disappear. The point at which this happens is known as its Curie temperature (after Pierre Curie, Marie's husband). For a material to be ferromagnetic, the magnetic spin axes of all its atoms need to be aligned. However, in gadolinium, thermal motion of the atoms disrupts this alignment, leading to potential applications in magnetically controlled refrigeration systems.

Gadolinium also responds strongly to the magnetic fields generated inside a magnetic resonance imaging (MRI) scanner. Injecting a patient with a gadolinium compound, formulated to be absorbed by a certain part of the body, enhances the image of that body part in an MRI scan.

Gd
64

Like many of the lanthanides, gadolinium is strongly
magnetic—until heated above 66 °F (19 °C), at which
temperature its magnetism abruptly disappears.
Gadolinium is added to other metals to improve
their corrosion resistance and workability.

Terbium

Category: lanthanide
Atomic number: 65

Atomic weight: 158.92535
Color: silvery white
Phase: solid

Melting point: 2,473 °F (1,356 °C)
Boiling point: 5,846 °F (3,230 °C)
Crystal structure: hexagonal

A number of chemical elements have been discovered in the rich mineral deposits near the village of Ytterby, in east central Sweden. Terbium is one of them and—like erbium, yttrium, and ytterbium—it, too, takes its name from the village.

Terbium was discovered by the Swedish chemist Carl Gustav Mosander, working in Stockholm. Mosander was studying samples of "yttria" (or yttrium oxide, to use its chemical name) gathered from a quarry just outside Ytterby. His investigations led him to suspect that the samples were harboring small quantities of another, unknown element. In 1843, Mosander was proved right when he succeeded in extracting terbium oxide, or "terbia," from the yttria.

Pure terbium is silvery white in color. It's malleable (easily squashed), ductile (easily stretched), and soft enough to be cut with a knife. The element forms part of the lanthanide sequence, a set of 15 elements with atomic numbers between 57 (lanthanum) and 71 (lutetium). Elements in the sequence were sometimes known as the "rare earths," but this title is now obsolete; they are not especially rare.

Terbium itself is quite reactive and so doesn't occur naturally in its pure metallic state. Instead, it exists within compounds— predominantly terbium oxide—locked away in minerals such as euxenite (1 percent terbium), monazite (0.03 percent terbium), and bastnasite (0.02 percent terbium). Another source is in clays found in southern China—these contain 1 percent terbium. Around 11 tons (10 tonnes) of the pure metal are produced every year from raw material gathered at sites in the USA, Sri Lanka, Australia, India, and Brazil—as well as China —and terbium is an expensive commodity, about four times the price of platinum .

Terbium, or rather an alloy of it known as Terfenol-D, expands and contracts under the application of a magnetic field—a curious property known as "magnetostriction." This means that wrapping an electromagnetic coil around a piece of Terfenol and passing a time-varying electric current through it—of the sort that might be used to encode an audio signal, for example—causes the Terfenol to vibrate in step with the current. If the Terfenol is then placed on a solid surface such as a tabletop, then the vibrations are transmitted to the surface, turning it into a giant loudspeaker. The effect has been exploited to develop the (now commercially available) Soundbug™ device, which can be attached to any resonant surface, such as a window pane, to turn it into a speaker for home audio systems. Terfenol-D is also used in naval sonar and is expected to play a role in developing motors and actuators for future microrobots.

Terbium is used to manufacture green phosphors for cathode-ray television tubes and fluorescent lighting, as well as solid-state electronic devices and lasers. It's also found an application in medical X-ray machines, where it's used to improve the sensitivity of the phosphors on the imaging screen. This has allowed the dosage of X-rays administered to patients to be reduced by a factor of four.

Tb 65

A lump of pure terbium metal. Terbium is a key ingredient in an alloy called Terfenol-D, which has the odd property of vibrating in time to any magnetic field that's applied to it. A piece of Terfenol-D can turn any object it's attached to into a loudspeaker.

Dysprosium

Category: lanthanide
Atomic number: 66

Atomic weight: 162.5
Color: silvery white
Phase: solid

Melting point: 2,565 °F (1,407 °C)
Boiling point: 4,653 °F (2,562 °C)
Crystal structure: hexagonal

The chemical element dysprosium was discovered in 1886 by the French chemist Paul Émile Lecoq de Boisbaudran. He found that samples of the element erbium seemed to be harboring some kind of unknown contaminants. These were later isolated as the elements holmium and thulium, in experiments carried out by Boisbaudran in the 1870s. He didn't stop there though, soon realizing that holmium oxide played host to the oxide of another unknown element. It took Boisbaudran an extremely protracted sequence of chemical steps to obtain a sample of this element. To mark the completion of such a Herculean task, Boisbaudran named his new chemical element "dysprosium," after the Greek *dysprositos,* meaning "hard to get."

And this was just to purify the oxide—making pure samples of the metal itself wouldn't be possible until the 1950s, when the Canadian chemist Frank Spedding, working at Iowa State University, perfected a technique known as "ion exchange chromatography." Pure dysprosium is a highly reactive metal that doesn't exist in nature. It is soft enough to be cut with a knife, corrodes rapidly in air, dissolves in acid, and gives off hydrogen on contact with water—meaning that no attempt should be made to extinguish burning dysprosium with water, lest an explosion result. There are seven naturally occurring isotopes of dysprosium, with atomic numbers 156, 158, and 160–164; none of them is radioactive.

Dysprosium forms part of the lanthanide sequence—15 chemical elements from lanthanum (atomic number 57) to lutetium (atomic number 71). Roughly 110 tons (100 tonnes) are produced annually, extracted from monazite, xenotime, gadolinite, and euxenite.

There are no definite estimates of global reserves of this element, but a shortfall is predicted; the price rose from US$7 per pound (US$15 per kg) in 2003 to more than US$127 per pound (US$280 per kg) by the end of 2010. The reason for this is that dysprosium is a minor constituent in the neodymium-iron-boron alloys used to make superpowerful magnets. These magnets are an essential component in new, high-efficiency electric motors, of the sort used in electric cars and other clean-energy technologies. Every hybrid electric car requires around 3.5 ounces (100 g) of dysprosium for its motor magnets. If millions of these cars are to be produced every year—as manufacturers hope—then hundreds of tons of the element will be required annually. For this reason, the United States Department of Energy predicts a global shortage of dysprosium by 2015—a fact that reinforces Boisbaudran's choice for this elusive element's name.

Other applications of dysprosium include neutron absorbers for nuclear reactors as well as in discharge lamps, where dysprosium compounds are used to produce very high-intensity light. It is also used in radioactivity dosimeters, where calcium crystals doped with dysprosium glow after exposure to ionizing radiation (such as-X rays and gamma rays)—the brightness of glow indicating the total dose received by the wearer.

Dysprosium is a soft, silvery metal. Its applications include high-power magnets, neutron-absorbing control rods in nuclear power plants, radiation dose meters, and high-intensity commercial lighting.

Holmium

Category: lanthanide
Atomic number: 67

Atomic weight: 164.93032
Color: silvery white
Phase: solid

Melting point: 2,662 °F (1,461 °C)
Boiling point: 4,928 °F (2,720 °C)
Crystal structure: hexagonal

Holmium is one of the lanthanides—a group of 15 chemical elements with similar properties set apart from the rest of the periodic table, and spanning the range of atomic numbers from 57 to 71. Whereas most chemical element families occupy vertical "groups" in the table, the lanthanides are unusual in that they align in a horizontal band.

All of the lanthanides have unusual magnetic properties, but holmium is exceptional in that it has the highest "magnetic moment" of any element. This is a measure that essentially quantifies the force a magnetic material is capable of exerting. When a material with a high magnetic moment is placed in a magnetic field, all of its atoms align themselves with the field, focusing and concentrating it. Of all the elements known, holmium does this best. For this reason, it is sometimes used to make so-called "pole pieces"—the lumps of metal sitting at the poles of a magnet.

This is particularly true in applications where a very strong magnetic field is required—such as inside an MRI scanner. An MRI scan works by subjecting the human body to an extremely strong magnetic field, which momentarily aligns all of its atoms. As the atoms snap back a short time later, they emit the magnetic energy as radiation, which is focused by the machine to create an internal image of the body.

Holmium was discovered in 1878 by two independent groups of researchers—Swiss chemists Jacques-Louis Soret and Marc Delafontaine, and the Swedish professor of chemistry Per Teodor Cleve. The sequence of discovery involves a number of the other lanthanide elements. During the early 1870s, the elements erbium and terbium were discovered lurking inside samples of yttrium. Later, ytterbium was found inside the erbium. It was in the residual material that the two groups of scientists found evidence for yet another new element. Soret and Delafontaine had done this by examining the spectrum of light given off by erbium, finding light at frequency bands that didn't fit with the element's known properties. This led them to speculate on the existence of a new element, which they named "Element X." When Cleve separated a new chemical from erbium, and its emission spectrum exactly matched with the mystery bands of Element X, he knew that he had found the elusive substance. He promptly gave it a new name: "holmium," after the Swedish capital Stockholm.

Holmium has just one naturally occurring isotope—Ho-165, which is stable. Other, synthetic isotopes exist, some of which are radioactive. The element oxidizes rapidly in the presence of moisture and cannot be found naturally in its pure form. Approximately 11 tons (10 tonnes) of holmium are extracted every year from deposits believed to total around 440,000 tons (400,000 tonnes).

In addition to its role in magnets, holmium is added to yttrium iron garnets to make eye-safe lasers that nevertheless have cutting power high enough for use in surgery, dentistry, and fiber-optic communications. Holmium's vivid red and yellow hues also make it useful as a colorant for cubic zirconia jewelry.

Ho

67

Holmium has the highest "magnetic moment" of any chemical element—meaning that its atoms readily align with and amplify magnetic fields. For this reason, it is used in some of the most powerful magnets in the world.

Erbium

Category: lanthanide
Atomic number: 68

Atomic weight: 167.259
Color: silver
Phase: solid

Melting point: 2,484 °F (1,362 °C)
Boiling point: 5,194 °F (2,868 °C)
Crystal structure: hexagonal

Erbium is another "rare earth" element, extracted from the gadolinium mineral deposits found near the village of Ytterby, in east central Sweden. Like some of the other rare earths, erbium derives its name from Ytterby, a village that, during the 19th century, became an epicenter for research into the fundamental chemical elements.

The rare earths form a sequence of 15 chemical elements with consecutive atomic numbers, starting with lanthanum (atomic number 57). For this reason, the modern name for the group is the "lanthanide sequence." Erbium itself was discovered in 1843 by Carl Gustav Mosander (see also terbium, page 152), who was the first to extract its oxide (erbia) from the Ytterby minerals; as with all the other lanthanides, its tendency to react with oxygen and water means that erbium doesn't exist in nature in its metallic pure form. Pure erbium metal wasn't produced until 1934, by reacting a sample of erbium chloride with potassium vapor. Erbium is one of the more abundant lanthanide elements, with a concentration in the Earth's crust of about 0.000045 ounces per pound (2.8 mg per kg).

The lanthanide sequence is a group of elements—all with very similar chemical properties—that is arranged horizontally along the bottom of the periodic table. This is due to a quirk in the way electrons are added to lanthanide elements as the atomic number increases. For example, erbium (atomic number 68) has one more electron than holmium (atomic number 67), but this electron is added to one of the inner electron shells. That means that the number of electrons in the outer shell—which is the principal factor determining an element's chemical properties—is the same for all the lanthanides: three.

Their chemical similarity tends to give the lanthanide elements similar physical properties. Many of them exhibit curious magnetic behavior, and are good substances to alloy with metals to reduce their hardness—thus rendering the metals more workable.

One exceptional application of erbium is in laser amplifiers for fiber-optic communication systems. Ordinarily, boosting the strength of the light pulse sent down an optical fiber means installing a complex relay that can convert the pulse into an electronic signal in order to amplify it. However, there is a much simpler alternative. Doping the glass in the fiber with erbium gives the fiber itself laser-like properties—causing it to emit photons of light of the same energy, and traveling in the same direction, as those in the original pulse.

The element also has uses in medicine. Erbium-based lasers give off infrared light with a wavelength of 2.9 micrometers. This is readily absorbed by water, making it perfect for surgical and dental applications, where a shallow depth of cut is essential.

Erbium is one of the few substances that can be used to make rose-tinted spectacles. Almost all pink glass will have been made so by doping with erbium oxide. The erbium oxide absorbs the green shades from white light, leaving the remainder a rosy pink.

Erbium is a soft silvery-white rare earth
metal. Its applications include amplifiers for
use in fiber-optic communications, infrared
lasers—and rose-tinted spectacles.

Thulium

Category: lanthanide
Atomic number: 69

Atomic weight: 168.93421
Color: silver–gray
Phase: solid

Melting point: 2,813 °F (1,545 °C)
Boiling point: 3,542 °F (1, 950 °C)
Crystal structure: hexagonal

Thulium was discovered in 1879 by the Swedish chemist Per Teodor Cleve, who was able to isolate it as an impurity in samples of erbium, an element that had been discovered just a few decades earlier. Cleve named his new element after Thule, an ancient term for lands at the boundary of the known world.

Thulium is a "rare earth" element, or "lanthanide," to use the more modern name. Due to a chemical quirk, all of the lanthanides have very similar physical properties—they are all soft, silvery metals that oxidize rapidly in the air and burn at relatively low temperatures: in the case of thulium, 300 °F (149 °C), much less than the 451 °F (233 °C) combustion point of paper. However, most of the lanthanides also have something special going on that distinguishes each element from the others. For example, erbium is extremely useful for amplifying fiber-optic communication signals, neodymium is used to make the most powerful magnets in existence, and gadolinium is an exceptional neutron absorber used to control activity in the cores of nuclear reactors.

Sadly, however, thulium has very little in the way of unique properties to make it stand out. To make matters worse, it's extremely rare and difficult to extract from base minerals—meaning that no one in their right mind would consider using thulium for its more generic properties when another (and much cheaper) lanthanide element would do the same job.

It's not completely bereft of applications, though. Lighting engineers use thulium in the construction of green-colored arc lamps—thulium is used as a coating on the bulb, where it absorbs energy from the spark inside the lamp and then reemits it at the characteristic wavelengths corresponding to green light. It's also used as a dopant (along with holmium and chromium) in YAG lasers, to create a material that gives off laser light highly suited to medical applications. Another application is in portable X-ray scanners; thulium that has been exposed to ionizing radiatin inside a nuclear reactor becomes a source of X-rays.

Thulium occurs in the Earth's crust at a meager concentration of 0.5 milligrams per kilogram. It is mainly found in the mineral monazite, of which it comprises just 0.007 percent. The so-called Oddo–Harkins rule explains this rarity. This states that elements with an odd atomic number (of which thulium is one) are always less abundant than their even-atomic-number neighbors.

The reason for this lies in quantum physics. Proton particles, the number of which in the nucleus of an element determines its atomic number, have a quantum property called "spin." This is quite different to spin in an everyday sense, in that quantum spin can only exist in one of two states, designated "up" and "down." Two protons with opposite spins tend to pair up to make a more stable configuration. The leftover particle in an odd-numbered group of protons is constantly trying to do the same thing. This creates a tendency for odd-numbered elements to increase their atomic number by one, and become even. Thus odd-numbered elements are rarer.

69
Tm

Odd-numbered element thulium is rare due to a
bizarre quirk of quantum physics. Its applications
include portable X-ray scanners, which can be used
in ambulances.

Ytterbium

Category: lanthanide
Atomic number: 70

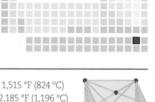

Atomic weight: 173.054
Color: silver
Phase: solid

Melting point: 1,515 °F (824 °C)
Boiling point: 2,185 °F (1,196 °C)
Crystal structure: face-centered cubic

As far as some chemists are concerned, ytterbium is the final element in the lanthanide sequence. The lanthanides form a series of chemical elements running from lanthanum in order of increasing atomic number. Historically, they were known as the "rare earths" and this designation also included the next element on from ytterbium, which is lutetium. Chemists disagree over whether lutetium is a lanthanide.

Ytterbium was discovered in 1878 by the Swiss chemist Jean Charles Galissard de Marignac, working at the University of Geneva. Ytterbium represents another step in a long and convoluted trail of element discovery involving the naturally occurring minerals found near the village of Ytterby, in east central Sweden. Many times that a new element was discovered from these minerals—and there have been many, including yttrium, erbium, terbium, and holmium—its samples turned out to be contaminated with substances that were new elements themselves. In 1843, the elements erbium and terbium were extracted from yttrium. Marignac was able to extract a white oxide from the erbium, which he correctly surmised to be the oxide of a new element. The element was named ytterbium, after the village of Ytterby. However, pure metallic ytterbium wasn't isolated from its oxide until the 1950s, using advanced ion-exchange techniques.

Nowadays, there are more plentiful sources of ytterbium. Probably the most abundant natural deposits are found in the ion adsorption clays of southern China, which contain around three to four percent ytterbium oxide. Found at an average concentration in the Earth's crust of around 3 milligrams per kilogram, ytterbium is twice as common as tin. Nevertheless, it is one of the scarcer lanthanides; we produce just 55 tons (50 tonnes) of pure ytterbium annually.

There are seven naturally occurring ytterbium isotopes, with mass numbers 168, 170–174, and 176. None is radioactive. However, the synthetic isotope Yb-169 is a strong emitter of gamma-rays—a form of ionizing radiation similar to X-rays. This has led to its use in portable X-ray machines, which have applications in emergency medicine (such as in battlefield hospitals), and for inspecting and diagnosing faults in machinery too large to be placed in a conventional scanner. It's also added to stainless steel to improve the metal's strength, and serves as a dopant in laser media, improving the quality of the light they emit.

Elements broadly divide into three categories: conductors, insulators, and those that lie somewhere in between, known as semiconductors. Ytterbium seems to hop back and forth in this classification scheme. At normal pressure, it behaves as a conductor, but if you squash it at 16,000 times atmospheric pressure, it turns into a semiconductor. Keep on increasing the pressure and, by 39,000 atmospheres, its conductivity will have dropped by a factor of 10, before shooting back up to around 10 times its ordinary value on reaching 40,000 atmospheres. This property is exploited in the construction of pressure gauges designed to operate in extreme conditions, such as inside nuclear explosions.

Ytterbium is one of the lanthanide series of elements,
many of which were discovered in minerals found
in eastern Sweden. They put the Swedish village of
Ytterby firmly on the map.

Lutetium

Category: lanthanide
Atomic number: 71

Atomic weight: 174.9668
Color: silver
Phase: solid

Melting point: 3,006 °F (1,652 °C)
Boiling point: 6,156 °F (3,402 °C)
Crystal structure: hexagonal

Some chemists include lutetium in the lanthanide series, while others put it in the transition metals. The lanthanides are a horizontal sequence moved from the main body of the table and, for the sake of neatness, appended at the bottom. The square they were taken from is often left blank, or sometimes even marked with a copy of lanthanum. Some, however, argue that lutetium should occupy this square—and that the element should correspondingly be removed from the lanthanide sequence.

Lutetium is actually the hardest and the densest element in the lanthanide sequence. This isn't surprising because, having the highest atomic number, it contains the largest number of particles. However, there's another reason, too. As you step through the lanthanide sequence in order of increasing atomic number, the diameters of the atoms, paradoxically, get smaller. This effect is called "lanthanide contraction." It's caused by the fact that increasing the atomic number increases both the positive charge on the nucleus and the negative charge of the electron cloud surrounding it. This in turn ramps up the force of attraction between the two, causing the atom to shrink. In most atoms, electrons are added to the outermost orbital level, or "shell," as the atomic number rises. This places the new electrons as far as they can be from the positively charged nucleus, minimizing the force of attraction. However, in the lanthanides, new electrons are added to an inner shell, meaning that they feel the electromagnetic pull of the

nucleus far more strongly—and this is the cause of lanthanide contraction.

Lutetium was discovered in 1907 by the French chemist Georges Urbain. It was the final step in a long trail of elemental discovery that began with yttrium in 1787. Because the properties of the lanthanides are all so similar, yttrium was able to harbor all kinds of other new elements that went undetected for many years, such as erbium, terbium, thulium, and holmium. Eventually, chemists were able to extract these elemental stowaways in a string of follow-up discoveries that culminated with lutetium. Urbain nearly missed out on getting the credit, as Austrian chemist Carl Auer von Welsbach and American Charles James both detected the element at around the same time. James was uncertain about his findings and declined to publish. However, Welsbach was bullish and a bitter dispute over who had found lutetium first erupted between him and Urbain. Priority was finally awarded to Urbain in 1909.

For many years, lutetium was the most expensive chemical element in the entire periodic table, at one point fetching US$34,000 per pound (US$75,000 per kg)—six times the price of gold. More effective methods of isolating it from raw minerals have now dropped the price to a more modest US$4,500 per pound (US$10,000 per kg). Annual production stands at around 11 tons (10 tonnes).

Lutetium has medical applications as a detector material in positron emission tomography (PET) scanners, while the isotope Lu-176 is used in radiotherapy.

The lanthanide metal lutetium is important in positron emission tomography (PET scanning), which is a medical-imaging technique for producing three-dimensional pictures of the interior of the human body.

Hafnium

Category: transition metal
Atomic number: 72

Atomic weight: 178.49
Color: silver–gray
Phase: solid

Melting point: 4,051 °F (2,233 °C)
Boiling point: 8,317 °F (4,603 °C)
Crystal structure: hexagonal

Chemical elements divide broadly into those that are radioactive, and so decay over time, and those that are stable, the quantity of which remains constant. Of all the stable chemical elements—and there are 81 of them—hafnium was the second to last to be discovered.

The reason for this is that, chemically, hafnium is almost identical to the element directly above it in group 4 of the periodic table—namely, zirconium. Members of the same group in the periodic table often have similar properties—indeed, it's a defining feature of the groups themselves. But hafnium occupies period 6 of the table, meaning that it shares characteristics with the lanthanides. These elements exhibit a phenomenon known as "lanthanide contraction," meaning that atoms shrink in size as the particle number increases. This effect conspires to make an atom of hafnium roughly the same size as an atom of zirconium, so it is nigh on impossible to tell the two apart. Indeed, spotting the difference between zirconium and the new element hafnium was a puzzle that took the best scientists many years to solve.

In 1923, physicist Dirk Coster and chemist Georg von Hevesy, working in Denmark, finally succeeded in this Herculean chemical task. They were able to detect hafnium using X-ray techniques to distinguish the different spectral lines from the two elements. The pair were driven by predictions of the element's existence, originally made in 1869 by Dmitri Mendeleev and later reinforced in 1913, when English physicist Henry Moseley discovered that elements are defined by the electrical charge (measured in multiples of the charge on an electron) on their atomic nuclei—a quantity known as "atomic number." Moseley realized that it is this quantity that determines the order of the elements in the periodic table. Once this was established, it immediately became clear that there was a gap in the table at atomic number 72. Coster and von Hevesy had now found this missing element. They named it hafnium, after *Hafnia*—the Latin name for Copenhagen, where they worked.

Although hafnium and zirconium are almost identical chemically, their nuclear properties are poles apart. Zirconium is used as nuclear reactor piping because it is virtually transparent to neutron particles; hafnium, on the other hand, is one of the best neutron absorbers known. One of hafnium's principal applications is in reactor control rods. Neutrons are the particles that perpetuate a nuclear reaction, thus being able to absorb them inside the reactor is crucial in order to control it.

Like other members of group 4, hafnium is highly resistant to corrosion and heat. Its compounds and alloys have been used in a number of high-temperature applications, such as rocket engines and plasma cutters. The latter are able to slice through sheet steel, which they do by using an electric current to heat a point on the steel to its combustion temperature and then blasting compressed air on the point, fanning the flames to burn through. Hafnium is used in the cutting tip of this device. Meanwhile, tantalum hafnium carbide has the highest melting point of any compound—7,619 °F (4,215 °C).

72

Hf

Being almost identical to its neighboring elements, hafnium eluded the finest brains of the chemical community for many years before it was finally isolated. The element and its alloys are very heat-resistant, leading to applications in plasma cutters.

Tantalum

Category: transition metal
Atomic number: 73

Atomic weight: 180.9479
Color: silvery gray
Phase: solid

Melting point: 5,463 °F (3,017 °C)
Boiling point: 9,856 °F (5,458 °C)
Crystal structure: body-centered cubic

Tantalum is named after a mythological tale of cannibalism and human sacrifice. The architect of this macabre story was Tantalus, King of Sisyphus, who was invited to dine at the table of the gods at Mount Olympus but angered them by stealing their food. In an attempt to appease the gods, Tantalus offered his son Pelops as a sacrifice, sliced up and boiled, but they declined the hideous repast and condemned him to be tormented for eternity by hunger and thirst. Banished to the Underworld, he was forced to stand up to his neck in water, which receded when he tried to drink, surrounded by boughs heavy with ripe grapes which were raised when he tried to eat.

The element itself proved tantalizing to isolate, which is perhaps how it earned its name. Tantalum is very similar to the element niobium, which is named after Niobe, Tantalus' daughter. In 1802 Anders Gustav Ekeberg, a professor of chemistry at Uppsala University, Sweden, claimed to have discovered a new metal—but when the esteemed British chemist William Hyde Wollaston examined the mineral in which it was found, he declared it to be identical to niobium. Forty years passed before tantalum and niobium were separated by Heinrich Rose, thus proving that Ekeberg's instincts were right.

Tantalum is a soft, silvery, shiny metal in its pure form. It occurs mainly as the isotope Ta-181 (99.99 percent) and its other isotope Ta-180 is very weakly radioactive. It is highly resistant to corrosion due to the oxide film that forms on its surface.

This proves useful for coating other metals, which not only protects them from corrosion but also provides an insulating layer. For this reason, tantalum is used in tiny capacitors (electronic components consisting of two metal plates separated by an insulating layer, which are capable of storing electrical charge) in items including cell phones, radios, video games, and medical equipment. Tantalum can create particularly strong alloys with other metals, which are used in rocket nozzles, heat shields, and airplanes.

The element is nontoxic and can be tolerated by humans, which has led to its use in surgery. If you've ever broken a bone so badly that it has needed fastening back together, the bolt that held it may well have contained tantalum. It has also been used for wire in sutures and in cranial repair plates.

Tantalum has been a controversial element since around 2000, when it emerged that mining for one of its key ores, coltan, in the Democratic Republic of the Congo, was having a disastrous effect on local wildlife. In the Kahuzi-Biega National Park, home of the mountain gorilla subspecies, ground was cleared to make mining easier and this reduced the amount of food available for the gorillas. To make matters worse, miners who were a long way from food sources killed gorillas to provide "bush meat" for themselves and rebel armies. In 2010, the United Nations reported that gorillas may disappear from much of the Congo Basin by the 2020s unless action is taken.

Tantalum is used in the manufacture of
electrical components that are so compact
there could be no cell phones, tablet
computers, or laptops without them.

Tungsten

Category: transition metal
Atomic number: 74

Atomic weight: 183.84
Color: silvery white
Phase: solid

Melting point: 6,192 °F (3,422 °C)
Boiling point: 10,031 °F (5,555 °C)
Crystal structure: body-centered cubic

Mention tungsten and most people will think of lightbulbs. Although the old, inefficient types that use tungsten have mostly been replaced by fluorescent energy-saving bulbs, the advent of the "incandescent" lightbulb changed life for ever—mostly in a positive way. Before 1879, when the American scientist Thomas Edison developed a practical, long-lasting lightbulb, people were dependent on candles or gas lights, which restricted both work and leisure pursuits. Suddenly, one 60-watt bulb could generate the same amount of light as around 100 candles, while removing the need for smelly and potentially dangerous gas lights.

Prior to 1906, when the US General Electrical Company created a tungsten filament, lightbulb filaments included metal-coated carbon, material derived from bamboo, and carbon fiber developed from cotton. Initially, this new tungsten filament was very costly, but American physicist William D. Coolidge improved on its design and created one that was affordable and would outlast all other forms.

Tungsten is the strongest metal at high temperatures and it is also inexpensive, which is why it was so useful in filaments. However, although the tungsten in an incandescent bulb would be heated yellow-hot, only 10 percent of the energy used would be converted into light; the other 90 percent would become heat or infrared radiation and therefore be wasted. For this reason, tungsten bulbs have mostly been replaced with environmentally friendly bulbs, which can last up to 20 times longer than incandescent ones and greatly reduce

your household or workplace carbon footprint. Tungsten is, however, still used as the filament in halogen lamps, which use elements such as bromine and iodine to prevent the tungsten filament from degrading; these are much more energy efficient than old-style lightbulbs although still not the most environmentally friendly option.

Tungsten is a silvery white, mildly toxic metal, found mainly in the ores scheelite and wolframite (from where its chemical symbol "W" is derived). The old name for wolframite was *wolfram*, which means "wolf dirt" in German; tin miners gave this name to certain stones that interfered with the smelting of tin and produced a lot more slag. To the miners, the tin was being "devoured" like a wolf would devour its prey.

The term tungsten is derived from the Swedish meaning "heavy stone," which used to be the name for the ore scheelite. In 1781, Carl Wilhelm Scheele studied the mineral tungsten and found that a new acidic white oxide could be made from it. Scheele deduced that this might be the oxide of a new element, but left off his research at that point.

In 1783, the Spanish brothers Juan José and Fausto Elhuyar studied the tungsten ore and wolframite and showed that both minerals yielded the same acidic metal oxide, which they were able to reduce to the metal we know as tungsten by heating with carbon. Fausto wanted to call the new element wolfram, but Juan José chose tungsten, after the name by which the metal ore was better known.

Pellets of silver-white tungsten metal. Tungsten
is the heaviest metal element to have a known
biological role, inside some prokaryote organisms
(single-celled life forms without a nucleus inside
the cell).

Tungsten can be brittle, due to impurities, but when heated in its pure state it becomes so ductile that it can be drawn into very fine wire. It has the highest melting point of all unalloyed metal and has the highest tensile strength at temperatures in excess of 3,000 °F (1,649 °C). Tungsten also has an excellent resistance to attack from oxygen, acids, and alkalis. World production of the metal in 2010 was around 75,800 tons (68,800 tonnes), most of which was mined in China. Tungsten can also be recycled, which accounts for almost 30 percent of the total amount of the element that is used in the USA.

While tungsten is something of a "workhorse" element, it has proved both useful and deadly in equal measure. When added to iron, it produces a particularly strong type of steel, which has been used to devastating effect in modern times. It all began in the 1860s, when British scientist Robert Mushet experimented by adding five percent tungsten to steel; he produced a metal so tough that it could withstand heat high enough to make the metal glow red. The alloy was used for machine tools, and when Germany recognized what other uses it could have, the nation began stockpiling reserves of it in the run-up to the First World War. The toughened steel was used in German cannons and guns, which had far superior firepower to French or Russian weaponry. Supply could barely meet demand and mines in neutral Portugal became the key sources of the tungsten that contributed to the carnage on the Western Front.

Tungsten is still used in armaments today, sintered with iron–nickel or iron–copper, which has particularly high density and is used for high-velocity antitank shells.

Today, one of the most important uses for tungsten is in cemented carbide, which is also known rather unimaginatively as "hard metal." This is made mainly from tungsten carbide, which is produced by heating tungsten powder and carbon powder in a furnace to almost 4,000 °F (2,200 °C). Cemented carbide is strong enough to cut cast iron and is used for tools essential in mining (including oil drills), metalworking (such as circular saws), and the construction industries. Tungsten carbide can also be found in very high-speed dental drills and on ballpoint pen tips, darts, fishing sinkers, and golf clubs. Tungsten carbide accounts for about 60 percent of the current global tungsten consumption.

The density and strength of tungsten make it the ideal choice for making heavy metal alloys. Its high melting point has led to its use in nozzles on rockets, including the submarine-launched Polaris ballistic missiles. Tungsten's high density has also made it useful in weights and counterweights, such as the ballast of yacht keels and Formula One racing cars and in the tail ballast of aircraft.

Tungsten has also been used in recent times in jewelry, where it is prized for its hardness and scratch resistance. A tungsten wedding ring is considerably cheaper than a gold or platinum one and is also hypoallergenic.

Tungsten lightbulb filaments enabled the world to see by electric light. The element is ideal for making filaments, because it can be drawn into very fine wire and heated to very high temperatures without melting.

Rhenium

Category: transition metal
Atomic number: 75

Atomic weight: 186.207
Color: silvery white
Phase: solid

Melting point: 5,767 °F (3,186 °C)
Boiling point: 10,105 °F (5,596 °C)
Crystal structure: hexagonal

Rhenium is a silvery white metal that is distinctive in having one of the highest melting points of all the elements—apart from tungsten and carbon. It is also particularly dense, with a close-packed hexagonal structure, and was the last stable element of the periodic table to be discovered.

At the turn of the 20th century, there were still some gaps in group 7 of the periodic table—although, from its position, number 75 was predicted to be a metal of high density with a range of oxides. The group is headed by manganese, which had been discovered at this point, and scientists attempted to isolate the missing element from manganese ores without much luck. It was finally discovered in 1925 by Walter Noddack and Ida Tacke in Berlin, who isolated it from the ore gadolinite, in which it was found in trace amounts. Otto Berg examined the concentrate they produced in a spectroscope and found several new lines that could only belong to a new element. Noddack and Tacke were able to obtain rhenium from molybdenite, although they could only extract an ounce of the element from 41,570 pounds (1 g from 660 kg) of ore. The new element was named after *Rhenus*, the Latin name for the river Rhine.

Rhenium is one of the rarest elements on Earth and does not occur as a free metal (most commercially used rhenium is produced in powder form). As a result, it is not known how rhenium would influence humans, plants, animals, or the environment. The element resists corrosion, but will tarnish slowly in moist air. There are two naturally occurring isotopes—Re-187, which is weakly radioactive, and Re-137, which is not. Re-187 has a half-life of 45 billion years; therefore most of the element that existed when Earth was formed, around 4.5 billion years ago, is still around us.

It was years before commercial quantities of rhenium could be produced and it is expensive because of its rarity. Due to its high melting point, it is used in superalloys—metals that have excellent mechanical strength at high temperatures and good corrosion resistance. Rhenium is used in nickel-iron alloys to make turbine blades in fighter-jet engines, which uses 75 percent of the global production of rhenium each year. Rhenium is also added to molybdenum and tungsten to make alloys that are suitable for use in filaments for lamps and heating elements in ovens.

Rhenium has also been used in thermocouples, which are devices that consist of two conductors (often metal alloys) that produce a voltage proportional to a temperature difference. Thermocouples can measure temperatures above 3,630 °F (1,999 °C) and are widely used in industry for gas turbine exhausts, diesel engines, and in kilns. The element is also used as a catalyst in hydrogenation reactions and is particularly useful because, unlike other catalysts, it is not deactivated by traces of phosphorus and sulfur. Hydrogenation reactions are frequently used to reduce or saturate complex organic compounds in the food, petrochemical, and agricultural industries.

Re 75

Having one of the highest melting points of all elements, rhenium is mostly combined with nickel and tungsten to form corrosion-resistant alloys that have excellent strength, even at very high temperatures.

Osmium

Category: transition metal
Atomic number: 76

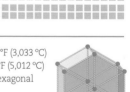

Atomic weight: 190.23
Color: bluish gray
Phase: solid

Melting point: 5,491 °F (3,033 °C)
Boiling point: 9,054 °F (5,012 °C)
Crystal structure: hexagonal

Osmium isn't one of the most renowned chemical elements, but it does have a few claims to fame: not only is it the densest of all elements, it is also the hardest pure metal found on Earth. Look closely and you'll see a delicate bluish tint to it, which also distinguishes it from the plethora of silver–gray metals on the periodic table.

The element exists in its natural form and also as alloys with iridium, which are called iridosmine or osmiridium. Pure osmium is very rare, however, and most is manufactured as a by-product of nickel refining. It is one of the platinum group of metals and shares a number of their properties, being hard, very dense, and durable.

Osmium was discovered in 1803 by the British chemist Smithson Tennant, who experimented by dissolving crude platinum in *aqua regia* (a particularly potent blend of acids). Other chemists had found that not all of the metal went into solution and assumed the powder that remained was graphite. Tennant was not convinced, though, and, through a series of treatments with acids and alkalis, he was able to separate the powder into two previously unknown elements: iridium and osmium. Tennant chose the name osmium because of the odor that his new-found element gave off (*osme* is the Greek word for "smell").

Osmium the element is not toxic, but its oxide is poisonous. Powdered osmium metal will gradually oxidize in air to form volatile osmium tetroxide, which will sublimate at room temperature to become a highly toxic vapor. Breathing in air containing just 0.10 micrograms per cubic meter of osmium tetroxide can cause irritation to the eyes, skin, and lungs and inhaling higher concentrations can lead to blindness and death.

At one time osmium tetroxide was used in fingerprint detection, because the vapor would react to traces of oils left by fingerprints to form a black deposit of a less harmful oxide. It is also used as a stain for electron microscopy, enhancing contrast by blocking the electron beam in parts of the sample.

Osmium has seen a number of uses over the years. In the late 19th century, the Austrian chemist Karl Auer created the Oslamp, which had a filament made of osmium; he introduced this commercially in 1902. However, osmium was soon replaced by the more stable tungsten, which became the standard for lightbulb filaments for most of the 20th century, until it was recognized how inefficient they were in terms of energy usage.

Like other members of the platinum group, osmium is mainly used in alloys where strength and durability are required. The styluses, or needles, of early phonographs (record players) used to be made of osmium alloy, as have the nibs of fountain pens, electrical contacts, and instrument pivots. One platinum-osmium alloy (which is 10 percent osmium) is used in pacemakers and replacement valves.

Osmium has seven naturally occurring isotopes, of which Os-192 is the most abundant. The isotope rhenium-187 decays to become osmium-187 and osmium-rhenium dating is useful in the study of meteorites.

Osmium is a member of the platinum group and shares the characteristics of its fellow elements, being resistant to chemical attack and having great strength and durability.

Iridium

Category: transition metal
Atomic number: 77

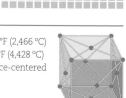

Atomic weight: 192.217
Color: silvery white
Phase: solid

Melting point: 4,471 °F (2,466 °C)
Boiling point: 8,002 °F (4,428 °C)
Crystal structure: face-centered cubic

Iridium is thought to have been part of a global cataclysm 65 million years ago, which led to the extinction of the dinosaurs and many other creatures on Earth. Geologists discovered its presence in rocks deposited when this devastating event occurred, at the cusp of the Cretaceous and Paleogene periods. There are a number of theories regarding what actually happened, but the most prevalent is that a meteor around 6 miles (10 km) wide blasted into the Yucatan peninsula, off the Gulf of Mexico, leaving a crater around 186 miles (300 km) in width. The resultant explosion sparked vast fires and sent clouds of dust into the atmosphere, which darkened the sky for so long that plants and creatures could not survive.

Iridium is relatively rare on Earth, but is abundant in meteorites. In 1980, Nobel Prize-winning physicist Luis Alvarez, his son Walter Alvarez, and nuclear chemists Frank Asaro and Helen V. Michael from the University of California, Berkeley, found unusually high amounts of iridium in a layer of rock strata in Earth's crust formed 65 million years ago. This geological signature has become known as the K–Pg boundary (K being the abbreviation for the Cretaceous period, and Pg for the Paleogene period). The layer is part of the fossil record and is visible at certain places, including Badlands, near Alberta, Canada, and at Stevns, on the island of Zealand in south Denmark. Two other meteorite impacts that occurred 36 million years ago in Chesapeake Bay in the USA and

Popigai, northern Russia, also left deposits of iridium, but these were not believed to have caused any mass extinctions.

Iridium was discovered at the same time as osmium, in 1803 by the British chemist Smithson Tennant. Like other chemists before him, he had dissolved crude platinum in *aqua regia* (a particularly potent mix of acids) and noticed that a black residue remained. Some believed this to be graphite, but Tennant wasn't entirely sure and began experimenting on the residue with alkalis and acids. The result of his efforts was the discovery of two previously unknown elements. He named one osmium because of its peculiar odor and the other iridium after Iris, the Greek goddess of the rainbow, because its salts were so colorful.

In its pure form iridium is brittle, lustrous, and silvery. It can be found in the iridium-osmium alloys, osmiridium, and iridosmine, although most of the element is obtained as a by-product of platinum refining.

The high melting point, corrosion resistance, and hardness of iridium and its alloys have determined most of its applications. In the past, iridium alloy was used for coating the nibs of fountain pens and in compass bearings. In modern times, it has a range of uses: to tip spark plugs in automobiles, to produce long-life aircraft engine parts, and to make crucibles that must withstand high temperatures. The radioactive isotope Ir-192 is a gamma-ray emitter and is used in radiation therapy for cancer patients.

Iridium played a crucial role in confirming the theory that the dinosaurs were wiped out by a giant asteroid from space. It is rare on Earth, but plentiful in asteroids—and plentiful in the layer of the geological record corresponding to the time when the dinosaurs died.

Platinum

Category: transition metal
Atomic number: 78

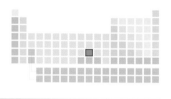

Atomic weight: 195.078
Color: silvery white
Phase: solid

Melting point: 3,215 °F (1,768 °C)
Boiling point: 6,917 °F (3,825 °C)
Crystal structure: face-centered cubic

Platinum is one of the rarest elements on Earth, which is one reason for the exclusivity associated with what is essentially another silvery white metal. After all, a platinum disk awarded to bestsellers in the music industry and a platinum credit card both hold more kudos than their gold equivalents.

Platinum leads its own "subgroup" on the periodic table, including ruthenium, rhodium, palladium, and iridium, which share similar chemical properties and often occur in the same mineral deposits. They are highly resistant to wear and tarnish, are excellent catalysts, and have good resistance to chemical attack. Platinum will not oxidize in air at all, which is why it has proved so popular in jewelry-making.

The earliest known specimen of worked platinum has been dated to the 7th century BC and was found in a trove dedicated to the Egyptian Queen Shapenapit, at Thebes; no other platinum artefacts have been found from ancient Egypt or Roman times, however. It is thought that South American people knew how to work platinum 2,000 years ago and deposits were particularly rich around the river Pinto in Columbia. The conquering Spaniards were, of course, very interested in gold, but they also became fascinated by a metal that looked much like silver but didn't tarnish, which they found alongside silver in the River Pinto. They even referred to it as *platina* ("little silver"), which gave us the element's modern name.

Platinum occurs as the pure metal, in ores and as an alloy of platinum and iridium. It is sourced chiefly in South Africa as cooperite

(platinum sulfide) and in the USA, where sperrylite (platinum arsenide) occurs in nickel deposits. The key uses of platinum are in jewelry and in catalytic converters, where it converts harmful unburned hydrocarbons in car exhaust fumes into CO_2 and water.

By accident, platinum played a role in the discovery of a hugely important chemotherapy drug called cisplatin. This drug has proved particularly effective against cancers of the testicles, bladder, lung, stomach, and ovaries. It was discovered in 1962 by the American chemist Barnett Rosenberg, who was carrying out experiments to determine if electromagnetic energy could stop cell growth. He was using platinum electrodes and noticed a dramatic effect on the *E. coli* cells he was monitoring: some of them grew to be 300 times their normal length and yet none of them divided. Rosenberg deduced that the platinum electrodes had reacted with chloride and ammonium ions in the culture medium to form a platinum compound: cis-diamminedichloroplatinum (II). This became the foundation of the drug cisplatin, which works by latching onto DNA at a certain point, distorting it in the process and preventing it from replicating. Early tests showed that the drug was highly effective in reducing tumors and, in 1978, it was approved for use in the USA. Cisplatin leapt to the forefront of cancer treatment and has achieved a 90 percent success rate overall. It's impossible to calculate how many lives it has saved.

Pictured is a platinum nugget weighing 13 lb (5.918 kg)—worth, to its lucky owner, more than US$300,000 (at 2013 prices). Platinum is not only a precious metal, but it is also an essential component of automotive catalytic converters.

Gold

Category: transition metal
Atomic number: 79

Atomic weight: 196.96655
Color: metallic yellow
Phase: solid

Melting point: 1,948 °F (1,064 °C)
Boiling point: 5,173 °F (2,856 °C)
Crystal structure: face-centered cubic

Gold's seductive beauty may have led people to risk life, limb, and long prison sentences in its pursuit, but there is another reason why it is so special: its chemistry. Gold is one of the least reactive metals, meaning that—unlike fellow transition metals, silver, iron, and copper—it doesn't tarnish. Gold is impervious to air, water, alkalis, and most acids, which is the reason why the gold used in the tombs of the ancient Egyptian pharaohs has retained its luster for millennia. Much of the element's symbolism is derived from this stability and longevity—hence we have golden wedding rings and celebrate golden wedding anniversaries (after 50 years of marriage).

Gold is only affected by *aqua regia* (a highly corrosive mix of acids), selenic acid, and water heated under pressure to 705 °F (374 °C). The reason behind its imperviousness is the firm grip that it has on its outermost electron (the one that is available to bond with other chemicals); the large positive charge of the nucleus has a powerful effect on its electrons, holding them in place.

The fact that gold is one of the least reactive metals is the reason why it is often found in its elemental form, as nuggets, alluvial deposits, and grains in rock. The largest nugget of gold was found in Australia and weighed in at 258 pounds (112 kg).

Gold is one of the most ductile (easily stretched) and malleable (easily squashed) metals; a single gram of it can be beaten into a one-meter squared sheet and gold leaf can be flattened until it becomes translucent. Gold will readily form alloys with copper, nickel, aluminum, and cobalt to provide myriad colors in jewelry.

The name gold is thought to be Anglo-Saxon and this may be derived from *geolo*, which means "yellow." Its chemical symbol, Au, comes from the Latin word *aurum*, meaning "glow of sunrise." In Italian, the word for gold is *oro*, and one of Venice's finest Grand Canal palazzi was called the Ca' d'Oro due to its ostentatious gold leaf external décor.

Gold was one of the first metals to be used in prehistoric times. Artefacts have been found dating from around 4700 BC in the Varna Necropolis ("City of the Dead"), in Bulgaria, one of the world's most important archaeological sites. Excavations in 1972 unearthed a network of almost 300 graves and in some an astonishing array of gold artefacts were found, denoting the high status or military rank of the owners. Finds included jewelry, ritual axes, animal figurines, and even a gold penis sheath.

Gold artefacts were also discovered in the royal graves of the ancient city state of Ur (in modern-day Iraq) and the early Egyptians found a plentiful source of alluvial gold in the River Nile. In 2008, a necklace made of turquoise and gold was found at Jiskairumoko, near Lake Titicaca in the Peruvian Andes; this was dated at 4,000 years old, making it the oldest native gold artefact to be found thus far in either of the American continents.

Today, gold is mined mostly in South Africa, Russia, the USA, Canada, Peru, and Australia. Some of South Africa's gold mines

This gold nugget was found in the Yukon River in 1841. Gold nuggets are brought together by the fluvial action of water, making rivers a productive hunting ground for prospectors. The largest nugget in history was the "Welcome Stranger," found in 1869 at Moliagul, Victoria, Australia. It weighed 172 lb (78 kg) and, at 2013 prices, would be worth almost US$4 million.

are more than 2.2 miles (3.5 km) deep. World production of gold is around 2,755 tons (2,500 tonnes) per year and it is used mainly in bullion, jewelry, electronics, and decorative glass. In recent years, India has become the world's largest single consumer of gold, purchasing 20 percent of all stock, mostly for jewelry.

Gold is also now used as a treatment for rheumatoid arthritis, when nonsteroidal anti-inflammatory drugs do not work. Injections of a weak solution containing compounds of gold—sodium aurothiomalate or aurothioglucose—are given to the patient and have proved effective in relieving joint pain and stiffness. The treatment cannot be used indefinitely, however, due to side effects that occur when the amount of gold in the body rises.

Gold is perhaps more commonly used in dentistry and this is not a modern phenomenon. Excavations in areas of Italy once inhabited by the Etruscans (around 900–300 BC) have revealed skulls with false teeth held together by gold dentistry. Around 66 tons (60 tonnes) of gold are used each year in dentistry; the gold used is hardened by being alloyed with palladium and silver.

In recent times, gold has been valued as an investment choice, particularly during times of recession; this has led to some calls for gold to be restored as a form of currency. The big flaw in the argument is gold's scarcity, however. If you were to gather together all the gold mined, it would fill a cube with sides measuring 60 feet (18 m)

wide. The total value of the gold stock ever mined would amount to around 10 trillion US dollars—one trillion dollars less than the total public debt in the USA in October 2012.

The oceans are a tantalizing source of gold, containing more than 11 million tons (10 million tonnes), with an estimated value of around 1,500 trillion US dollars. Seawater in general contains 11 parts per trillion of gold; however, the cost and challenges of deep-sea gold mining make extracting it economically nonviable at the moment. This has not deterred teams of bounty hunters who are prepared to risk perilous conditions in search of treasure. The Bering Sea has some of the highest concentrations of gold, perhaps due to the gold-rich rivers of Alaska and Siberia that flow into it, and gold hunters are risking their lives by diving through pack ice to the bottom of the sea, where they use high-powered vacuums to try to harvest the precious metal. Use of deep-sea robots to navigate the ocean floor and surveying advances could improve the chances of finding treasure in the deep, but conservationists argue that mining could destroy sea life and harm the natural harmony of the marine environment.

It's an intriguing thought that, thanks to its longevity, this beautiful element may outlive the human race. From the mask of King Tut to the wedding ring on your finger, gold will be one of the most enduring elements and one that tells future generations (or invaders) how the human race lived, loved, fought, and died.

Gold bars are the medium used by the world's central banks to store their monetary reserves. As of spring 2013, gold was priced at over US$50,000 per kg. A single "standard" gold bar (like those pictured) weighs 12.4 kg, and is thus currently worth more than US$620,000. Gold increased in value almost six-fold between 2000 and 2013.

Mercury

Category: transition metal
Atomic number: 80

Atomic weight: 200.59
Color: silvery white
Phase: liquid

Melting point: -38 °F (-39 °C)
Boiling point: 674 °F (357 °C)
Crystal structure: rhombohedral

Mercury has inspired a sense of mystique for centuries; early alchemists believed it to be the "first matter," from which all other metals were formed. This theory was dispelled in the mid-18th century when two Russian scientists, A. Braun and M.V. Lomonosov, experimented with snow and acids to achieve very low temperatures. To their surprise, when they attempted to take a reading, the mercury in their thermometer stopped moving and appeared to have solidified. They broke the glass and found that the mercury in the bulb had become a solid metal ball and the mercury in the tube could be bent, like other metals. The mystique was beginning to fade.

Mercury's chemical symbol Hg is derived from the Latin word *hydragyrum*, meaning "liquid silver," and the element also has the common name of quicksilver. Mercury is the only metal that is liquid at standard temperature and pressure and it is so dense that even lead will float on it. The oldest known sample of pure mercury was discovered in an Egyptian tomb dating to 1600 BC, but the bright red compound mercury sulfide, or cinnabar, was used 30,000 years ago to decorate the walls of caves in Spain and France. Mercury itself is a rare element in Earth's crust, but mercury ores are extremely concentrated.

For thousands of years, mercury was believed to prolong life and boost general health, in particular in China and Tibet. Of course, it had the opposite effect. Ancient Greeks used mercury in ointments and Egyptians and Romans used it in cosmetics,

which would sometimes result in facial deformities. Through history, doctors used calomel (mercury chloride) as a laxative, a diuretic, and to treat venereal disease. It is believed that England's Henry VIII and Russia's Ivan the Terrible were both treated with mercury for syphilis.

Hair includes sulfur-containing amino acids that attract mercury, and is therefore a good indication of a person's exposure to the element. English scientist Sir Isaac Newton's hair was found to contain high levels of mercury, although he was celibate for almost all of his life, so this was more likely to have resulted from alchemy than syphilis.

Sadly, it took millennia before people understood that mercury was toxic and that exposure to it would damage the brain and central nervous system and lead to madness. The phrase "mad as a hatter" was coined to describe the behavior of hatmakers whose job it was to dip fur pelts in mercury so that they would mat together. The industry terms "hatter's shakes" or "mercury madness" describe the effect the chemical had on the workers over time. Early 20th-century detectives searching for fingerprints at the scene of a crime were also at risk of mercury poisoning, because the dusting powder they used was made of chalk and mercury ground together.

Mercury is ideal for thermometers and barometers, because it moves as pressure and temperature dictate. However, due to its toxicity, it has been largely replaced by digital thermometers. Mercury is still used in mercury-vapor lights and fluorescent lamps.

The highly toxic chemical element mercury is the
only metal that is liquid at room temperature. It
remains liquid right down to -38 °F (-39 °C) and
all the way up to 674 °F (357 °C)—well beyond the
points at which most other liquids have either
frozen or boiled.

Thallium

Category: post-transition metal
Atomic number: 81

Atomic weight: 204.3833
Color: silvery white
Phase: solid

Melting point: 579 °F (304 °C)
Boiling point: 2,683 °F (1,473 °C)
Crystal structure: hexagonal

Thallium has enabled some people to get away with murder, literally. This is one of the most toxic elements on the periodic table and, like arsenic, was once dubbed "inheritance powder." Even today, investigators may not suspect thallium poisoning because its symptoms—vomiting, delirium, hair loss, and abdominal pain—can be confused with other afflictions.

The pure element thallium is a soft, silvery metal that tarnishes in damp air and is attacked by acids. It is fairly abundant on Earth and occurs mostly in the potassium mineral sylvite and the cesium mineral pollucite. It was discovered in 1861 by British chemist William Crookes of the Royal College of Science in London, using spectroscopy. He was examining some sulfuric acid and observed a green line in the spectrum that had not been noticed before. Crookes realized that he had found a new element and named it thallium after the Greek word *thallos*, meaning "green twig."

Controversy loomed around the corner, however. In 1862, Claude-Auguste Lamy, of Lille in France, researched thallium more intensively and managed to extract the pure metal and create an ingot. He sent this impressive specimen to the London International Exhibition, where committee members declared it to be a new metal and awarded Lamy a medal. Crookes was furious and a cross-Channel war of words began until Crookes was placated with the award of his own medal. But now, let's go back to poisoning and murder …

After it became easy to test for arsenic in 1836, thallium became the "poisoner's poison," mainly because it was easy to administer and difficult to detect. The poison was usually given in the form of thallium sulfate, which dissolves in water and is tasteless, colorless, and odorless. At one time, the salt was readily available over the counter of stores, as a rat or ant poison, and a dose of 0.0176 ounces (500 mg) would usually prove fatal.

In 1953, Australians were horrified to discover that they had a female serial killer in their midst. Plump, jolly grandmother Caroline Grills loved to treat her family to delicious homebaked goodies; however, in 1947 a number of them began to fall ill and die. First her 87-year-old stepmother passed away, then three in-laws, including her sister-in-law, died in a similar way. A relative witnessed Grills attempting to pour something into a cup of tea for another member of the family and he was able to switch the cups and keep the suspicious beverage. Police tested it for thallium and the result came back positive. Grills was arrested and charged with four murders and three attempted murders. Doubts were raised by the defense over Grills' mental health, although she did stand to gain financially from the killings. She was found guilty and sentenced to death, which was commuted to life imprisonment. Grills became known affectionately as "Aunt Thally" by the inmates of Sydney Long Bay prison, where she remained until her death in 1960. Thallium sulfate is banned in most nations now, although it is still used as a pesticide in some developing countries.

Thallium makes cocktails with a kick. It's highly toxic and difficult to detect, and so is perfect for expediting the demise of unwanted family members. At least, that was Australian serial killer Caroline Grills' reasoning when she used the stuff to kill four relatives.

Lead

Category: post-transition metal
Atomic number: 82

Atomic weight: 207.2
Color: gray
Phase: solid

Melting point: 621 °F (327 °C)
Boiling point: 3,180 °F (1,749 °C)
Crystal structure: face-centered cubic

In 1845, the explorer Sir John Franklin set sail with two ships to try and seek out the northwest passage—a sea route through the Arctic Ocean. Neither he, nor any of his crew, was ever seen again—at least, not alive. The graves of three crew members were later discovered on Beechey Island; analysis revealed that they had died from lead poisoning. Franklin's expedition had packed enough food to last five years and much of this was canned—then a relatively new technique for preserving food. The technique involved packing the cans with food through a small hole and then soldering a piece of metal over the hole to form an airtight seal. The trouble was that the solder was made with lead—Franklin's crew had been poisoned by their rations, the very thing that was meant to sustain them throughout their arduous trek.

Humans have been using lead for thousands of years. Ancient Egyptian tombs have been found to contain cosmetics made with lead compounds such as galena (lead sulfide) and cerussite (lead carbonate). One compound found in the tombs, phosgenite, is extremely rare in nature, making it likely that the Egyptians had actually learned to manufacture it. However, because lead has been with us for so long, the name of its discoverer is not known.

One of the earliest big users of lead was the Roman empire, thought to have mined 88,000 tons (80,000) tonnes of the stuff per year. They excavated it principally for its silver content—there's 2.6 pounds (1.2 kg) of silver in every ton of lead ore—and the Romans used much of their lead for piping. It was the ideal material for the development of a running water system—easily worked and rolled into sheets, which could then be formed into tubes and welded to make them watertight. Welding was easy, owing to the melting point of lead—sufficiently low to make liquid metal on a simple wood fire. The chemical symbol, Pb, derives from the Latin name for lead, *plumbum*. The word, together with the principal application of lead in pipework, is what gives us the modern word for people who work with piping: plumbers.

As well as tipping molten cauldrons of the stuff over anyone who dared attack their fortifications, the Romans were so fond of lead that they even used it to sweeten wine. No wonder their fertility levels eventually began to suffer from the slow lead poisoning of the population. Lead poisoning also causes damage to the brain, as well as anemia.

Lead has found many applications since— many of them now abandoned following concerns over toxicity. In 1921, American engineer Thomas Midgley discovered that adding a lead compound to gasoline raised its octane level. It was later to bring untold levels of pollution, as evidenced by the amount of lead deposited from the atmosphere in Arctic snow. Virtually the whole world has now switched to unleaded fuel. Paint, fishing weights, and pottery glazes all went the same way. Some useful applications remain, though: lead's corrosion resistance and workability make it an excellent material for roofing houses; and without lead radiation shields, X-ray machines and other medical scanners would not be possible.

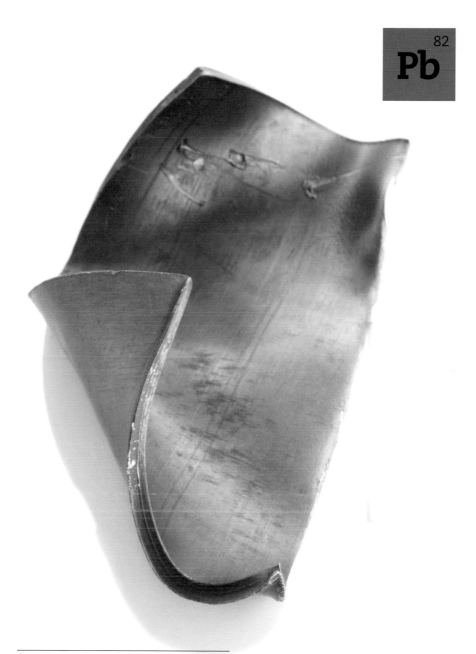

Lead is positioned amid a cluster of poisonous elements, such as thallium, mercury, and polonium. Its toxicity has led to it being replaced in certain applications (namely those in which killing people is not an objective) by its neighbor bismuth, which has low toxicity.

Bismuth

Category: post-transition metal
Atomic number: 83

Atomic weight: 208.98040
Color: silver
Phase: solid

Melting point: 521 °F (272 °C)
Boiling point: 2,847 °F (1,564 °C)
Crystal structure: trigonal

Bismuth is one of those elements that have been known about for donkey's years. It's thought to have been first discovered in the early 15th century by a nameless practitioner of alchemy. Alchemy was a pseudoscientific forerunner to the true science of chemistry—its aim being to somehow convert plentiful metals, such as lead, into gold. This is possible but, sadly for the alchemists, requires a nuclear reactor—the likes of which wouldn't be seen for another 500 years.

Bismuth is roughly twice as abundant as gold in the Earth's crust. It is heavy (86 percent the density of lead) and is a silver metal with a faint pink tinge. On contact with air, bismuth oxidizes to form an attractive iridescent film over the element's surface. There is one main naturally occurring isotope of bismuth, with atomic number 209. This isotope is very weakly radioactive, with a half-life of 1.9×10^{19} years.

Bismuth is produced from mined ore, extracted from sites in locations including Peru, Bolivia, and Japan. Most, however, comes as a by-product of the refining of other ores—for example, copper and tin smelting.

Pure bismuth is a very brittle substance, but that doesn't stop it from having a number of useful applications when alloyed with other metals. One of the earliest of these was invented around 1450, when the extremely low melting point of lead-bismuth alloy led to its use as a typesetting material in printing. "Hot metal" typesetting, as it's known, literally involves injecting molten alloy into molds, to form pieces of type called "sorts," which can then be assembled to make printing plates.

By the late 15th century, bismuth alloys had found other uses in ornamental metalwork. Nevertheless, many confused the element with its close neighbors on the periodic table, lead and tin. The distinction wasn't made clear until 1753, when French chemist Claude François Geoffroy demonstrated their differences.

The origin of the name bismuth isn't entirely clear. One theory says that it comes from the German *wismuth*, itself derived from *weisse masse* ("white mass")—which could refer to the white compound bismuth oxychloride. Another holds that the word is descended from *bi ismid*, an Arabic term meaning "similar to antimony."

Since the 18th century, bismuth has been used in the treatment of stomach complaints such as peptic ulcers. It's still used today as a medication for diarrhea. This is somewhat surprising if you consider that bismuth's direct neighbors on the periodic table are lead, an element whose reputation for toxicity goes before it, and polonium, an element so poisonous that it's been used in at least one assassination, probably more.

Applications for bismuth are many and varied, including cosmetics—bismuth oxychloride is used to produce the lustrous pearlescent finish of some white nail varnishes. It's used to make hunting ammunition that won't poison the environment like lead does. Bismuth vanadate forms a vivid yellow pigment for paint, replacing the toxic lead chromate used previously. Bismuth is also used in pyrotechnics to make those loud, crackly fireworks that are the staple of any display.

Bi
83

Bismuth is trivalent (meaning each atom forms
strong bonds with up to three others). Compared
to its neighbors on the periodic table, bismuth is
remarkably nontoxic and is even used in some oral
medications for stomach upsets.

Polonium

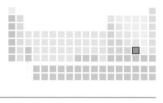

Category: metalloid
Atomic number: 84

Atomic weight: 209
Color: silvery gray
Phase: solid

Melting point: 489 °F (254 °C)
Boiling point: 1,764 °F (962 °C)
Crystal structure: cubic

One of the deadliest substances known to man, polonium is a trillion times more toxic than hydrogen cyanide. Polonium is so radioactive that a solid lump of its most common isotope Po-210, will glow due to the excitation of the air surrounding it.

The element was discovered in 1898 by French physicists Marie and Pierre Curie, who noticed that the uranium pitchblende that they were studying was even more radioactive after the removal of uranium and thorium. This spurred the Curies on to find additional elements and they isolated polonium first before radium. The element is named after Poland, where Marie was born.

Uranium ores contain minute amounts of polonium—this element is usually made in a nuclear reactor by bombarding bismuth with neutrons. Despite its deadly nature, the element does see some uses outside nuclear facilities. It is used in antistatic brushes, where a thin layer of polonium is sandwiched between gold and silver on a foil mounted just behind the bristles. The polonium ionizes the air, conducting the static charge away. The polonium is made *in situ* by plating the silver foil with bismuth and then gold and running the foil under a neutron beam that converts the bismuth to polonium. Thus, the polonium never exists in the open, where it would present huge danger given that just 10 nanograms of it can be fatal.

The darker side of this element gave rise to tragedy in 2006, when Russian dissident and ex-KGB officer Alexander Litvinenko died from polonium-210 poisoning in London. Litvinenko was a vocal opponent of Russian premier Vladimir Putin and publicly accused the Putin regime of ordering the assassination of Russian journalist and human rights activist Anna Politkovskaya. On November 1, 2006, Litvinenko met two Russian men, including Andrei Lugovoi, an ex-KGB officer, at the Millennium Hotel in Mayfair. That night, he fell ill with stomach pains and vomiting, which continued for three days before he was admitted to hospital.

Litvinenko suffered a painful, public, and lingering death, passing away 23 days after his fateful meeting. Following tests, the UK Health Agency revealed that there were significant amounts of polonium-210 in Litvinenko's body—a substance rarely found anywhere outside the laboratory. Experts believed that less than a microgram (one millionth of a gram) of polonium killed Litvinenko. A higher dose would have killed him far more quickly.

Fear began to spread as a "polonium trail" emerged across London, with traces of the isotope being found in the hotel, a sushi bar, and a club Litvinenko visited that evening. Most of the contaminated locations were cleared of danger in the weeks following Litvinenko's death. An inquest into the incident was ordered in 2011—though only after tireless campaigning by Litvinenko's widow, Marina.

Other victims of polonium poisoning include Marie and Pierre Curie's daughter Irène Joliot-Curie, who died in 1956. Irène was also a Nobel Prize-winning scientist, and she died of leukemia, thought to have been caused by exposure to polonium after a capsule exploded on her bench 15 years earlier.

A sample of pitchblende, a form of the uranium
ore uraninite. It was from this substance that the
French chemists Pierre and Marie Curie were first
able to isolate polonium in 1898.

Astatine

Category: halogen
Atomic number: 85

Atomic weight: 210
Phase: solid
Color: unknown

Melting point: 576 °F (302 °C)
Boiling point: 639 °F (337 °C)
Crystal structure: unknown

Astatine is one of the rarest elements on Earth, which is not a bad thing considering the fact that it is highly radioactive. It completes the halogen group and shares some of the characteristics of its fellow elements: it is highly unstable in its natural form (its name is derived from the Greek word *astatos*, which means "unstable") and, like iodine, it will vaporize quickly at room temperature.

Astatine is heavier than other halogens, which is what makes its melting and boiling points higher than the rest of the group. Quantities of astatine on Earth are estimated at just 28 gm at any one time and the half-life of its longest lasting isotope, astatine-210, is just eight hours. Such characteristics create huge challenges for the research chemist— as a result, the pure element has never been seen by the naked eye.

In 1869, when Mendeleev published his periodic table, there was a space under iodine and after the Danish physicist Niels Bohr established the classification of the elements, it was suggested that there might well be a fifth halogen. For a time this was known as "eka-iodine," from the Sanskrit *eka* meaning "one," referring to the element one space under iodine. Chemists attempted to find this elusive element in nature, without any luck; until, it seemed, 1931, when a group of scientists led by Fred Allison in Alabama, USA, claimed that they had isolated eka-iodine. They gave it the atomic number 85 and called it alabamine. However, their claim was disputed after other scientists failed to recreate their processes successfully.

The element was finally isolated in 1940 by a team at the University of California, Berkeley, USA, including Dale R. Corson, Kenneth Ross Mackenzie, and Emilio Segrè, an Italian physicist who had escaped from fascist Italy. They isolated astatine by bombarding bismuth with alpha particles in a device called a cyclotron, a form of particle accelerator. However, the team could not continue its research due to the Second World War and the demands of the Manhattan Project (the US government research project that developed the atomic bomb). Since then, research into astatine and its properties has been difficult due to its highly reactive nature.

Only tiny amounts of astatine exist in nature as part of the decay of uranium (page 210) and thorium (page 206), and the element has no biological role. The isotope astatine-211 has found use in recent times in radiation therapy for cancer, in particular prostate cancer and melanomas. However, once the astatine is produced, it has to be used very quickly due to its short half-life. Like iodine, astatine is collected by the thyroid gland in the neck.

Astatine was once thought to be the rarest element on Earth; however, it has lost this status to berkelium (page 222), which has to be produced by neutron capture and beta decay in uranium.

Autunite, a fluorescent uranium mineral whose decay can produce an atom or two of astatine.

Radon

Category: noble gas
Atomic number: 86

Atomic weight: 222
Color: colorless
Phase: gas

Melting point: -96 °F (-71 °C)
Boiling point: -79 °F (-62 °C)
Crystal structure: n/a

It's a chilling thought that a highly radioactive gas could collect in the basement of your house, but it's true; this danger can occur because radon gas forms through the natural decay of uranium and thorium in granite bedrock and diffuses from there into the atmosphere. The amount of radon in air varies from place to place and we all breathe in a tiny amount each day, but if it collects in a confined space you have a problem. Radon is thought to be the major contributor to our daily exposure to background radiation, and a serious health hazard to miners.

Radon is believed to cause between 3 and 14 percent of all lung cancers. It is the leading cause of lung cancer among nonsmokers (although smokers are still more at risk).

Radon gas was first noted by the German chemist Friedrich Ernst Dorn, who was investigating the pressure of gas created by radium compounds. He did not claim the discovery, however. In 1902, the eminent British nuclear physicist Ernest Rutherford and his assistant Frederick Soddy studied radioactive gas emanating from thorium and declared that they had found a new element. Six years later, their compatriots William Ramsay, who had discovered four other inert gases, and chemist Robert Whytlaw-Gray, noted that it was the heaviest gas and named it *niton*, from the Latin meaning "shining" (which it did in the dark). In 1923, the International Committee of Chemical Elements agreed on the name radon, which is derived from radium.

Radon is an inert gas, which places it in a group with other chemically harmless gases. However, the danger that radon presents is down to its radioactive isotopes, known as "radon daughters," which are fine, solid particles, formed by the element's radioactive decay, that emit harmful alpha radiation. If inhaled over time, these particles can damage the sensitive lung tissue and lead to cancer.

Radon can gather in buildings made out of granite (for example, Grand Central Station, New York) and seep up to collect in rooms with poor ventilation. During the 1980s and 1990s, this risk became apparent and radon gas home detector kits were produced. Those at risk were advised to seal cracks in concrete floors and walls, and to ensure effective ventilation. In serious cases, a space was created below the floor where the radon could collect and be extracted. In most homes, exposure to radon is around 20 becquerels (the unit of radioactivity) per cubic meter. In uranium mines, this can rise to 10 million becquerels per cubic meter.

Before the harmful effects of radiation were discovered, there was a craze for all things radioactive, with many people believing that radioactive substances could actually boost health. Some would take radioactive quack medical products and flock to thermal spas that emitted radiation. Surprisingly, some people still believe that inhaling radon can be a good thing for their health and journey to the spa town of Le Mont-Dore in Auvergne, France, to partake of the radon-rich air. They do this through a process known as "nasal irrigation," where a tube is inserted up a nostril so that they can inhale the gas direct from the spring.

RADON BULBS

For the preparation of
Radio - Active Water
in a Sparklet Syphon

Box of Six - 3/-
Further supplies
in exchange for
empty bulbs   Box of Six - 2/6

Sole Makers:
SPARKLETS LIMITED.
Head Office: Thames House, Millbank, Westminster, S.W.1.
Works: Upper Edmonton, London, N.18.

In the early 20th century, it was thought that radon water—water infused with radioactive radon gas—was the ultimate health drink. These capsules of radon gas were used in the household to produce the stuff. However, radon water was anything but healthy.

Francium

Category: alkali metal
Atomic number: 87

Atomic weight: 223
(isotope francium-223)
Phase: solid
Color: unknown

Melting point: 80 °F (27 °C)
(estimated)
Boiling point: 1,256 °F (680 °C)
(estimated)
Crystal structure: body-centered cubic

Francium is intensely radioactive and is one of the least stable elements in the periodic table. It occurs naturally in uranium minerals but only in the tiniest quantities, as a product of the radioactive decay of actinium-227. Fr-223, the element's longest-living isotope, has a half-life of only 22 minutes and at any one time it is estimated that there is only 1.05 ounces (30 g) of the element in existence on Earth.

Francium is created for research purposes only, in a particle accelerator by bombarding thorium with protons, or in a nuclear reactor by bombarding radium with neutrons. Due to its very short half-life, francium has no uses commercially or in medicine, which has made use of other highly radioactive isotopes instead.

Research into francium is a challenge, because it is so radioactive that a chunk of it would probably evaporate swiftly due to the heat that it would produce. Throw any alkali metal into water and you're guaranteed a dramatic display—but if you were brave/ crazy enough to introduce francium to water, the explosion would be massive. The element is assumed to possess similar chemical and physical attributes to cesium, which occupies the position above it on the periodic table. Francium has 30 isotopes, but very little is known about them because most have half-lives of less than one minute.

The element was discovered in 1939 by the chemist Marguerite Perey, who worked at the Curie Institute in Paris. Mendeleev, when he created the periodic table, deduced that there must be another element below cesium, which he dubbed eka-cesium (*eka* is the Sanskrit for "one"). As more refined versions of the table were produced, other chemists confirmed Mendeleev's hypothesis; a number reported that they had found the elusive element and gave it names including russium, virginium, and moldavium. Their claims were all disproved.

Perey was born in 1909 and showed an aptitude for science from an early age. She joined the Radium Institute in Paris in 1929 as a laboratory assistant and was mentored by Marie Curie, who quickly recognized talent in the young scientist. The pair worked together until Marie Curie died of leukemia in 1934, and Perey continued their studies, focusing primarily on the radioactive decay of the element actinium. This was to result in her most important discovery.

After producing a sample of actinium, free of radioactive decay products, Perey was intrigued to see that the emission of beta radiation was more intense than it ought to have been. She concluded that there must be an unknown element present and hypothesized that it was element 87. She was proved right and named the new element after her homeland.

Perey was honored by becoming the first female member of the 200-year-old Institut de France, Académie des Sciences. Not even the great Marie Curie had broken through the gender barrier to achieve this feat. Perey also became professor of nuclear physics at the University of Strasbourg. She died of cancer in 1975, like so many other pioneering researchers into the radioactive elements.

Francium is radioactive, unstable, and reacts insatiably with other chemicals. Today it only remains in trace quantities on Earth. You might find a few atoms of it as decay products in samples of the uranium ore pitchblende— like the one pictured.

Radium

Category: alkaline earth metal
Atomic number: 88

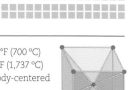

Atomic weight: 226
Color: white
Phase: solid

Melting point: 1,292 °F (700 °C)
Boiling point: 3,159 °F (1,737 °C)
Crystal structure: body-centered cubic

Radium is a soft, radioactive metal that was discovered in 1898 by Pierre and Marie Curie; they extracted it from uranium pitchblende. It was named radium from the Latin word *radius,* for its power of emitting energy rays.

Marie clearly found her discovery captivating and said, "one of our joys was to go into our workroom at night when we perceived the feebly luminous silhouettes of the bottles and capsules. ... the glowing tubes looked like faint fairy lights." Radium's luminescence is caused by its radioactivity; the nucleus of each atom is unstable and breaks apart over time, giving off energy. All isotopes of radium are highly radioactive. Although in its pure form, it is a pristine silvery white metal, radium readily oxidizes in air and grows dark in color.

Radium quickly became regarded as a "miracle" cure or tonic in the early years of the 20th century, with unscrupulous "quacks" peddling a range of radium-related "wonder drugs." These ranged from a "tonic" known as Raithor to radium suppositories and the eyewatering radienocrinator, which was designed to be worn by men over the groin to promote health and virility. Of course, all had quite the opposite effect.

For many years at the start of the 20th century, radium was a key part of the treatment of cancer. "Radium needles" would be inserted into tumors and the cancerous cells would be blasted with the intense radiation they emitted. The dangers of radium soon became apparent, however. Workers who prepared the radium needles had low white blood cell counts and some died of radium exposure. In 1900, reports emerged of the first case of "radium dermatitis," and after Marie Curie experimented by carrying a tiny sample in contact with her skin for 10 hours, it was just a few days before she developed an ulcer.

Radium became useful in the manufacture of luminous alarm clocks and watches, the dials of which were hand painted. This intricate work was done predominantly by young girls, who would lick their brushes to get a fine point. The paint contained small amounts of radium, but it was enough to cause cancerous growths among a number of workers. This led to the famous case in the 1920s of the five American "radium girls," who took their employer US Radium to trial as a result of their fatal illnesses. The company tried to wriggle out of its responsibilities but eventually settled out of court, awarding each girl US$10,000. Within a few years, however, all five were dead.

Some of the girls at US Radium were so contaminated that their faces and hair glowed in the dark. Subsequently, workers producing luminous dials for aircraft, gun sights, and electrical instruments were provided with a much safer working environment, with screens, ventilation, and intensive cleaning processes.

Marie Curie became another victim of exposure to radioactive elements when she died in 1934 due to aplastic anemia. This is a condition that prevents the bone marrow from producing enough blood cells. The Curies' papers from the 1890s are too dangerous to handle and even Marie's cookbook is highly radioactive.

This is the material that radium is made from. It's a sample of black uraninite (uranium oxide)—along with a deposit of yellowish-brown gummite (uranium hydroxide). Sharp-cornered black crystals of uraninite are occasionally found in nature, but it usually occurs in lustrous, black masses (as here), when it is known as pitchblende.

Actinium

Category: actinide
Atomic number: 89

Atomic weight: 227
Color: silver
Phase: solid

Melting point: 1,922 °F (1,050 °C)
Boiling point: 5,788 °F (3,198 °C)
Crystal structure: face-centered cubic

Actinium is the first element in the "actinide sequence," a horizontal strip of 15 elements extending to lawrencium (atomic number 103), moved out of the periodic table's main body for neatness and appended at the bottom—much like the lanthanides. However, whereas the lanthanides occupy period 6 of the table, the actinides occupy the row below, period 7. While the lanthanides all exhibit very similar chemical properties, the actinides are remarkably diverse; and whereas the lanthanides are as safe as houses, the actinides are all radioactive—most of them lethally so.

Actinium was discovered in 1899 by French chemist André-Louis Debierne, who was able to isolate the new element in experiments involving the same uranium pitchblende from which Marie and Pierre Curie had extracted radium. German chemist Friedrich Oskar Giesel independently found it in 1902. Even though there has been some debate over which scientist deserves the credit, most agree that Debierne got there first. The name actinium derives from the Greek word *aktis*, meaning "beam" or "ray"—a reference to the element's radioactivity.

Actinium is a soft-silvery metal, with similar mechanical properties to lead. In the dark, actinium glows a pale blue color, as its natural radioactivity knocks the electrons from air atoms—a process known as ionization. Like other elements in group 4 of the periodic table, it reacts with oxygen in the air to form a layer over its surface that guards against further oxidation. Actinium gives off hydrogen on contact with water.

There is just one naturally occurring actinium isotope—Ac-227, which has a radioactive half-life of 21.8 years. Thirty-six further isotopes have been manufactured synthetically—some with half-lives shorter than a few minutes. Actinium occurs naturally on Earth but only in tiny amounts, accounting for about 0.2 parts per billion of uranium ore. For this reason, extracting it from uranium is a difficult and costly business; most actinium is therefore synthesized, which is done by bombarding radium with neutron particles. Absorbing one neutron increases the mass of radium from 226 to 227, but this isotope of radium is unstable and then decays by beta emission to actinium. About 2 percent of the mass of a radium sample can be converted to actinium in this way.

Even despite our ability to manufacture it, the amount of actinium available for use is tiny—a fact reflected in the narrow range of applications it is put to. Clad in beryllium, it makes an effective neutron source for use in laboratory experiments. There has also been some interest in exploiting actinium's radioactivity in radioisotope thermoelectric generators—devices that generate electricity from radioactive heat, for use on long-duration spaceflight far from the Sun, where there is no appreciable solar energy. Actinium has also found an application in targeted radiotherapy, for eradicating tumors. However, the element accumulates in the bones and liver, so a "chelating agent" is later administered to mop up the toxic actinium, converting it into a form that is readily excreted.

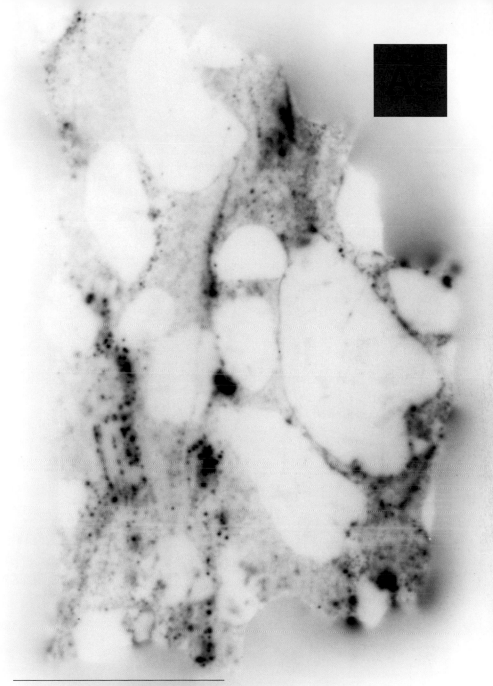

Traces of actinium can be isolated from pitchblende, a type of radioactive uranium ore. This image was produced by placing pitchblende on a sheet of sensitive X-ray film for four days. The darkest spots represent the highest sources of radiation.

Thorium

Category: actinide
Atomic number: 90

Atomic weight: 232.0381
Color: silver
Phase: solid

Melting point: 3,348 °F (1,842 °C)
Boiling point: 8,650 °F (4,788 °C)
Crystal structure: face-centered cubic

Thorium is one of the most abundant radioactive elements on Earth, being some 200 times more common than silver and three times more common than humble tin. Incredibly, for many years its abundance led many to overlook its danger, as a highly radioactive substance; thorium was in the past used in all sorts of everyday applications from gas mantles to camera lenses. Nowadays, the uses for thorium are more restricted but nonetheless important—with some scientists arguing that it could be a plentiful fuel source for the next generation of nuclear reactors.

Thorium was discovered in 1829, by Swedish chemist Jöns Jakob Berzelius, who named it after Thor, the Norse god of thunder. It wasn't the first time. In 1815, Berzelius isolated what he thought was a new element and named it thorium—only to discover that what he had actually found was yttrium, an element that had already been discovered in the late 18th century. He obviously liked the name so much that he used it again when he extracted a genuine new element from a rock found on Norway's Lovoy Island. Thorium's radioactivity was not discovered until 1898, thanks to the work of the mother of radioactivity, physicist Marie Curie.

Thorium was later responsible for the discovery of one of the fundamental properties of radioactive elements—half-life. In the early 20th century, physicist Ernest Rutherford and colleagues showed how thorium decays into other elements over a well-defined timescale, the half-life—the time for the radioactivity of a sample to have decayed by half. The finding supported the theory that radioactivity was caused by the spontaneous disintegration of atomic nuclei.

The most common isotope of thorium has an atomic mass of 232 and a half-life of 14 billion years. Given the age of the Earth (4.5 billion years), this means that 85 percent of primordial Th-232 is still here. There are a total of 27 isotopes known, but almost all naturally occurring thorium is in the form of Th-232.

Added to magnesium, thorium makes an alloy known as Mag-Thor, which is especially strong, lightweight, and resistant to temperature, and which was used in missiles and aircraft. Thorium also has applications in arc lights and arc welding equipment, where a small amount of thorium added to the tip, or "button," of the arc generator increases its melting temperature and improves the quality of the arc.

Whereas today, judicious use of radioactive isotopes brings benefits to medicine via radiotherapy, in years gone by the dangers were not understood. Thorium-based liquids were injected into patients to enhance X-ray images, while so-called thorium "health drinks" would bring anything but.

However, radioactivity has its upside, too. Thorium is a potential fuel source for nuclear fission reactors—especially given the fact that it is three times as common in the Earth's crust as uranium, the fuel used by most commercial reactors today. Some scientists even argue that a reactor based on thorium could be inherently safer than current designs, and would produce less dangerous waste.

This incandescent gas mantle consists of fabric impregnated with an oxide of thorium. When heated from within by a flame, the fabric will glow with a white light. The thorium is absorbing infrared heat energy and then reemitting it as visible light.

Protactinium

Category: actinide
Atomic number: 91

Atomic weight: 231.03588
Color: silver
Phase: solid

Melting point: 2,854 °F (1,568 °C)
Boiling point: 7,280 °F (4,027 °C)
Crystal structure: tetragonal

Protactinium is a radioactive chemical element belonging to the actinide sequence—a group of heavy elements starting at atomic number 89 and extending up to 103. The actinides are appended below the main body of the periodic table as a horizontal strip.

In 1871, the inventor of the periodic table, Russian chemist Dmitri Mendeleev, was the first to suggest that there might be an unknown element lurking between thorium and uranium. In Mendeleev's original version of the table there was no space for an actinide sequence—because the quantum theory describing the orbits of electrons, from which the actinide sequence follows, was yet to be discovered. And so Mendeleev placed the new element in the slot below tantalum.

The first sniff of Mendeleev's missing element was gleaned by British physicist William Crookes. He was able to isolate a new radioactive element from samples of uranium, which he imaginatively named uranium-X. In 1913, German physicists Kasimir Fajans and Otto Gohring showed that Crookes's uranium-X decays by beta emission—they named it brevium after the word "brevity," in reference to its short half-life (6 hours and 42 minutes).

Brevium was, in fact, an isotope of what would turn out to be protactinium—one with 231 particles in its atomic nucleus. In 1918, another two German physicists—Lise Meitner and Otto Hahn—were able to find a longer-lived isotope of protactinium, Pa-234 (half-life: 32,760 years), which they extracted from pitchblende (a uranium ore).

They published their findings later that year, along with Fajans and a team from the UK consisting of physicists Frederick Soddy, John Cranston, and Andrew Fleck.

In their paper, Meitner and Hahn proposed the name "protoactinium," from the Greek *protos*, meaning "first"—a reference to the fact that this element is a precursor to actinium (the decay of protactinium gradually converts its atoms into those of actinium). This name was accepted but, in 1949, was shortened to the more pronounceable "protactinium."

It's a miracle that protactinium was ever discovered. Mendeleev's initial classification of the element suggested it should have chemical properties similar to those of tantalum. We now know that it's not in the same group as this element and so there is no reason for any similarity. However, by sheer fluke, many of protactinium's properties did turn out to mirror those of tantalum. A total of 29 protactinium isotopes are now known, although only two of them occur naturally, Pa-231 and Pa-234.

Protactinium is extremely arduous to produce, accounting for just three parts per million in uranium ores. In 1961, the UK Atomic Energy Authority manufactured a batch of 99.9 percent pure protactinium, at a cost of US$500,000. From 66 tons (60 tonnes) of spent uranium fuel, they managed to wring just 4.4 ounces (125 g) of the element. Samples of this material were sent around the world for analysis. Perhaps due in part to its extreme rareness and toxicity, protactinium remains today an element that has no applications beyond its use in pure scientific research.

The mineral tobernite contains traces of
the actinide element protactinium. Despite
being largely useless, with no commercial
applications, in 1961 4.4 oz (125 g) of
protactinium was manufactured in the UK.
No one seems quite sure why.

Uranium

Category: actinide
Atomic number: 92

Atomic weight: 238.02891
Color: silver–gray
Phase: solid

Melting point: 2,070 °F (1,132 °C)
Boiling point: 7,468 °F (4,131 °C)
Crystal structure: orthorhombic

Uranium is arguably the most infamous of all the chemical elements. Radioactive, chemically toxic, it is the awesome power source that brings light and energy into most of our homes—but it is also the substance at the heart of some of the most terrifying weaponry that human beings have ever devised.

The first atomic bomb to be used in anger was code-named "Little Boy" and dropped on the Japanese city of Hiroshima at 8:16 a.m. on August 6, 1945. It contained 141 pounds (64 kg) of uranium, of which perhaps only 2.2 pounds (1 kg) actually detonated. Nevertheless, this was enough to release the same quantity of energy liberated by 13 kilotons, or 13,000 tonnes, of conventional high explosive. The resulting blast killed 75,000 people and leveled 50,000 buildings. Modern nuclear weapons make even this look feeble, though—their typical yields are measured in the hundreds and thousands of kilotons, and hold as much as 100 times the power of "Little Boy."

The Hiroshima bomb was a nuclear fission device, deriving its power by splitting apart the nuclei of uranium atoms. Fission is a result of German physicist Albert Einstein's formula $E=mc^2$, essentially stating that energy and mass are equivalent, related by a constant of proportionality equal to the speed of light squared. Physicists had noticed that the mass of the atomic nucleus divided by the number of particles it contains is bigger for heavy elements, suggesting that the nuclei of large, heavy elements (such as uranium) harbor excess energy that can be released—in huge amounts—when they are broken up into smaller pieces.

The splitting of atoms is triggered by the capture of slow-moving neutron particles. Because they are slow moving, the neutrons are absorbed by the nucleus rather than simply bouncing off. These extra particles then make the nucleus unstable, and ultimately cause it to break apart. Crucially, the splitting process creates more slow neutrons, which can repeat the process on other uranium nuclei, setting up a self-sustaining nuclear chain reaction.

This is the same process by which modern nuclear reactors work. Of course, we don't want a runaway chain reaction inside a nuclear power plant, and so reactors use a material called a "moderator" to filter the number of neutrons and stop the reaction from boiling over into a full-scale explosion—2.2 pounds (1 kg) of uranium fuel can provide as much energy as over 1,653 tons (1,500 tonnes) of coal.

Amazingly, nuclear reactors don't always need to be man made. There is evidence that, 1.7 billion years ago, rich natural uranium deposits beneath the ground at Oklo in Gabon, Africa, spontaneously ignited a nuclear chain reaction, moderated by natural running water.

At an average concentration of three parts per million, uranium is 40 times more abundant than silver in the Earth's crust. Total deposits there are estimated at 10^{17} kilograms, or 100,000 billion tonnes, and more than 58,000 tons (53,000 tonnes) are mined annually. Approximately 31 percent

A mixture of uraninite (black) and its chemical reaction products gummite (yellow) and curite (orange). Uraninite is an oxide of uranium, and is an important ore of the element. It is mostly mined in Congo, South Africa, Australia, and Canada.

of this comes from Australia, with the largest known concentration of uranium ore being at the Olympic Dam Mine, in South Australia.

The two main naturally occurring isotopes of uranium are U-235 and U-238. Small quantities of U-234 exist naturally and other isotopes, such as U-234 and U-239, have been manufactured artificially. Terrestrial uranium deposits are 99.3 percent U-238 and just 0.7 percent U-235. It's the scarcer U-235 that's needed for nuclear fission, meaning that before naturally mined uranium can be used as a nuclear fuel, it must first undergo a process known as "enrichment" to increase its concentration of U-235—usually to above three percent. When the natural reactor at Gabon was running, 1.7 billion years ago, naturally occurring concentrations of uranium-235 were already at this level, so no enrichment was needed. Today, levels are much lower, due to natural radioactive decay.

Uranium is weakly radioactive, giving off alpha particles. Of the naturally occurring isotopes, U-238 has the longest half-life, taking 4.468 billion years for its radioactivity to decay by 50 percent. In contrast, U-234 has the shortest half-life, at 248,000 years.

In addition, uranium is chemically toxic, with exposure causing serious damage to many regions of the body, including the kidneys and brain, and leading to an increased risk of cancer.

The radioactive decay of uranium is a major source of the Earth's internal heat, which in turn is responsible for volcanism and plate tectonics, the forces that make the Earth's crust shift and that cause earthquakes. Studies of uranium decay have helped provide estimates for the age of the Earth of around 4.57 billion years.

Uranium enrichment naturally creates a waste product with reduced U-235 content, typically no more than 0.3 percent: this is known as "depleted" uranium. This substance has found uses in applications that require extremely high-density materials (it is 70 percent denser than lead)—in particular, armor-piercing projectiles for destroying tanks. These are made all the more effective by the fact that uranium is exceptionally hard and that the pulverized uranium metal that penetrates the tank is pyrophoric, bursting into flames spontaneously upon contact with the air. Uranium's radioactivity, coupled to its chemical toxicity, mean that depleted uranium dust may explain the so-called "Gulf War Syndrome," as extensive use was made of depleted uranium ammunition during the conflict.

Incredibly, early applications of uranium included pigments for pottery and glass; the earliest known example of a uranium glaze comes from a first-century-AD Roman villa on the Bay of Naples, Italy. Orange-glazed "Fiestaware" was made and sold up until 1942, even though its radioactive emission could set off a Geiger counter (a device for detecting ionizing radiation) from a distance of a few feet. Needless to say, if you have any such items in your home you should probably remove them—and certainly not eat your dinner from them.

Uranium was discovered as such in 1789 by German chemist Martin Heinrich Klaproth, who was able to extract an oxide of the element from the radioactive ore pitchblende. He named it after Uranus, which had recently been discovered as the seventh planet of the Solar System. The silvery white uranium metal was first isolated in 1841 by French chemist Eugène-Melchior Peligot. However, its radioactivity wasn't discovered until even later still, in 1896, when French physicist Antoine Henri Becquerel noticed that a sample of uranium salt fogged a photographic plate kept in the same dark drawer. This one discovery led to the brand-new field of radiochemistry.

A "button" of uranium-235, the radioactive isotope that is used as a fuel for nuclear reactors and as an explosive in nuclear weapons, such as the Hiroshima bomb. It weighs about almost 10 lb (4.5 kg) and is worth approximately US$200,000.

Neptunium

Category: actinide
Atomic number: 93

Atomic weight: 237
Color: silver
Phase: solid

Melting point: 1,179 °F (637 °C)
Boiling point: 7,232 °F (4,000 °C)
Crystal structure: orthorhombic

Neptunium is a member of the actinide sequence (a run of radioactive elements with atomic numbers from 89 to 103) and is the last naturally occurring chemical element on Earth. (Elements with atomic numbers higher than 93 can only be manufactured in nuclear reactors and particle accelerators.)

Neptunium was a difficult element to discover because initially it was thought to occupy a spot *in* the periodic table (rather than appended below as part of the actinide sequence) and beneath rhenium; it was also thought, wrongly, to share chemical properties with this element. However, this was before the true structure of atoms was discovered (thanks to quantum theory), which in turn led to the discovery of the actinide (and lanthanide) sequences. Discovering the actinides effectively widened the table and made it clear that element 93 is more similar to its horizontal neighbors than to the elements above it.

Italian–American physicist Enrico Fermi was the first person to try and find element 93. In the 1930s, Fermi and colleagues tried bombarding uranium with neutrons—the idea being that a uranium nucleus could absorb a neutron, which would then decay into a proton, an electron, and a neutrino particle. The latter two particles would be emitted (as "beta radiation"), while the proton would raise the atomic number of the resulting element by one, thus creating a nucleus of a new element. Fermi failed dismally. His first two creations—which he named ausenium and herperium—turned out to be just fragments, created from the fission of uranium atoms, which he had inadvertently caused by bombarding the element with neutrons.

The real element 93 didn't make itself known until May 1940, when American Edwin McMillan and colleague Philip Abelson were able to create a genuine sample using Fermi's technique. Given it was the element next to uranium—named after the Solar System's seventh planet—they named the new element after the eighth, Neptune. McMillan shared the 1951 Nobel Prize in Chemistry for his work on the so-called transuranium elements (those heavier than uranium).

Physically, neptunium is a silver–gray metal that reacts readily with water as well as steam and acids. It has the highest liquid range of any element, with 6,053 °F (3,363 °C) separating its melting and boiling points.

Neptunium only turns up on Earth in minute quantities, usually within uranium ores. This is because neptunium is formed as part of the natural radioactive decay of uranium into plutonium. Today, most of the neptunium used is extracted from spent nuclear fuel rods. Ultimately, all of the naturally occurring uranium will complete the transition to plutonium and then there will be no neptunium at all left in the world.

Neptunium has limited uses as a detector of high-energy neutrons. Despite being such an exotic element, the chances are that most people have some neptunium in their homes. It is a decay product of the radioactive element americium, a tiny sample of which is used in household smoke detectors.

Neptunium exists naturally in tiny trace amounts inside certain radioactive minerals. For example, these pieces of radioactive eschynite probably harbor a few atoms of neptunium, created during the radioactive decay of other elements.

Plutonium

Category: actinide
Atomic number: 94

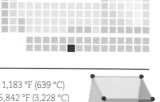

Atomic weight: 244
Color: silver–white
Phase: solid

Melting point: 1,183 °F (639 °C)
Boiling point: 5,842 °F (3,228 °C)
Crystal structure: monoclinic

At 11.02 a.m. on August 9, 1945, the Japanese city of Nagasaki was flattened by a sphere of metal about the size of an orange; an estimated 70,000 people died and 100,000 more were injured. The metal that wrought this devastation was called plutonium, and it was the central component in the second atomic bomb to be used in anger. The first had been dropped three days earlier, on the Japanese city of Hiroshima. This first bomb was a uranium device and it had required some 132 pounds (60 kg) of uranium; such was the potency of plutonium, that a mere 13.7 pounds (6.2 kg) of the stuff was enough to scratch Nagasaki from the map and force Japan's surrender six days later, ending the Second World War.

The Nagasaki plutonium bomb, code-named "Fat Man," was a radically different design to "Little Boy," the uranium-based Hiroshima device. Initially, it was thought possible to make a plutonium weapon that operated in the same way as Little Boy, by slamming together two pieces of nuclear fuel to create a "critical mass"—a lump big enough to set off a runaway nuclear chain reaction. But it soon became clear that plutonium gave off too many neutrons—the particles that trigger each nucleus to split apart and liberate energy. This meant that as soon as two pieces of plutonium came anywhere near each other, the reaction would start prematurely—leading to a so-called "fizzle," where the bomb fails to detonate correctly. The solution was to take a single subcritical piece of plutonium and surround it with conventional high explosive. Detonating the explosive compressed the fuel at the center, while at the same time the core was showered with neutrons from an "initiator" device made of beryllium and polonium. This scheme worked perfectly, as had been proven a month earlier in the Trinity test—the world's first ever nuclear detonation, which used a plutonium bomb of the same design.

Plutonium was discovered in 1940 at the University of California, Berkeley, USA, by a team of physicists and chemists led by Glenn Seaborg. The team bombarded uranium-238 with deuterons (heavy hydrogen nuclei, consisting of a proton and a neutron), which first formed neptunium, with atomic number 93. Neptunium then decayed by beta emission (raising the atomic number by one more) to form element 94. Seaborg chose to continue the trend of naming elements for planets and (since Pluto was still classified as a planet back then) called it plutonium.

There are 19 plutonium isotopes known, with the most fissionable of these being Pu-239. The longest-lived isotope is Pu-244, with a half-life of 80.8 million years. Pu-238 has found applications as a power source. It gives off mainly weakly penetrating alpha radiation, but enough of it so that a reasonable-sized lump can generate hundreds of watts of power for long periods. The Voyager spacecraft, as well as the Cassini probe to Saturn, incorporated such "radioisotope thermoelectric generators" (RTGs). Surgically implanted heart pacemakers now run on innocuous lithium batteries, but early models were powered by a small piece of radioactive plutonium.

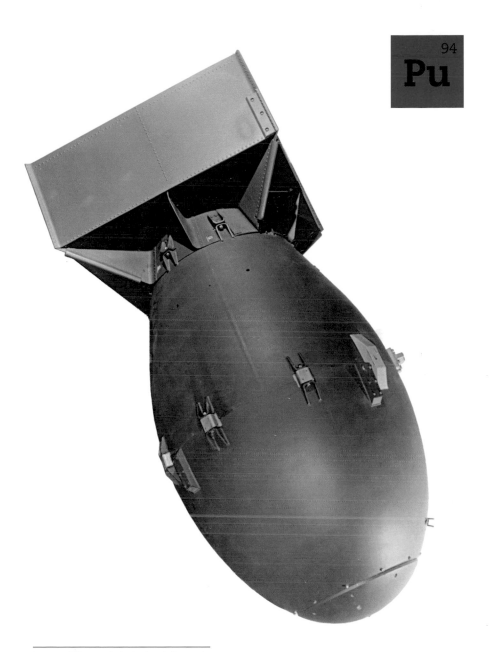

Pu 94

"Fat Man," the atomic fission bomb dropped over Nagasaki, Japan, on August 9, 1945. The bomb consisted of a slightly subcritical mass of plutonium encased by chemical high explosives. It was 10.8 ft (3.3 m) long and had a yield of 21 kilotons—equivalent to 23,000 tons (21,000 tonnes) of TNT.

Americium

Category: actinide
Atomic number: 95

Atomic weight: 243
Color: silver–white
Phase: solid

Melting point: 2,149 °F (1,176 °C)
Boiling point: 4,725 °F (2,607 °C)
Crystal structure: hexagonal

Americium is a chemical element with atomic number 95, placing it on the actinide sequence—a group of radioactive elements appended below the main body of the periodic table. Americium was discovered in 1944, by a team at the University of California, USA, led by pioneering American chemist Glenn Seaborg (see page 216). It was Seaborg himself who shaped the periodic table in its current form, introducing both the lanthanide and actinide sequences. The team used the university's 60-inch (152-cm) cyclotron, an early particle accelerator, to bombard atoms of plutonium with neutron particles. The nuclei of some of the atoms gained an extra neutron, which then decayed into a proton, raising the atoms' atomic number from 94 to 95. The name americium was chosen as a counterpoint to the element's opposite number in the lanthanide sequence—europium.

Americium was discovered as part of the USA's nuclear research during the Second World War and as such it was a closely guarded secret until the war was over—just in case it turned out to have important military applications. Its existence was due to be announced at a meeting of the American Chemical Society on November 11, 1945. However, on November 6, Seaborg appeared on a children's radio show, called *Quiz Kids*. When asked by an astute listener whether any new transuranium elements had been discovered during the war, Seaborg saw little point in lying just for the sake of five days.

There are 19 isotopes of americium; the most common are Am-241 and Am-243, with half-lives of 432.2 and 7,370 years,

respectively. Owing to its powerful radioactivity (it's more radioactive than weapons-grade plutonium), it doesn't occur naturally. It therefore must be manufactured either chemically or through nuclear reactions, as was done by Seaborg's team in 1944. One source is spent nuclear fuel, with a ton of reactor waste harboring around 3.5 ounces (100 g) of this element.

Americium's principal application is in smoke detectors. A typical household smoke detector contains a tiny sliver of Am-241 foil. The foil is radioactive, emitting alpha particles into a small chamber within the detector. These particles strip electrons from atoms and molecules in the air, making the air electrically conductive so that a current can flow through it. If smoke particles enter the chamber, they absorb some of the alpha particles, lowering the air's conductivity and causing the alarm to sound.

Some people have expressed concern upon discovering that they have such radioactive material in their homes. However, such a small quantity of americium is only a very weak source of alpha radiation. The number of lives saved by smoke detectors far outweighs any infinitesimal risk posed by the radioactivity. More worrying is that there are no special laws governing the disposal of smoke detectors. This led to the curious story of David Hahn, the "radioactive boy scout," who in 1994, having achieved his scouting badge in Atomic Energy, attempted to build a nuclear breeder reactor in his backyard using radioactive material largely from salvaged smoke detectors. Thankfully, the device never achieved critical mass.

A sample of radioactive americium. Interestingly, many of us aren't that far away from a piece of this chemical element. It is used in household smoke detectors—when smoke blocks the travel of radioactive particles between the americium source and a detector, the alarm is triggered.

Curium

Category: actinide
Atomic number: 96

Atomic weight: 247
Color: silver
Phase: solid

Melting point: 2,444 °F (1,340 °C)
Boiling point: 5,630 °F (3,110 °C)
Crystal structure: hexagonal close-packed

Curium is the first in a long line of elements at the heavy end of the periodic table that are named after people. It is also a member of the actinide sequence—a group of radioactive elements stretching horizontally across the periodic table from atomic number 89 to 103. The actinides were only discovered relatively recently. Together with the lanthanides, they form a group of elements normally removed from their usual location in the table and displayed at the bottom.

The element curium is a hard, dense, silver metal, but it wasn't immediately found in its metallic form. In fact, it wasn't immediately apparent in any form at all. Curium is highly radioactive. Any natural deposits that were present on planet Earth when it formed have long since decayed to nothing. That means that curium today must be manufactured inside nuclear reactors. This was first achieved in 1944 by a team at the University of California, Berkeley, USA, using the university's 60-inch (152-cm) cyclotron particle accelerator. The team, led by chemical element guru Glenn Seaborg, bombarded a sample of plutonium with alpha particles (essentially helium nuclei). Some of the particles stuck to the plutonium nuclei, raising their atomic number to 96.

The irradiated metal was subsequently passed to the university's metallurgical laboratory, which was able to infer the presence of the new element. The discovery, however, took place in 1944, while the Second World War was still raging. Since the element was found as part of America's nuclear weapons program, it was not announced until after the end of hostilities in 1945. The team named the element in honor of Marie and Pierre Curie, the French pioneers of radioactivity.

The first curium compound made in sufficient quantities to be visible to the naked eye was produced in 1947; first piece of pure curium metal was manufactured in 1951. Curium is now made predominantly by bombarding plutonium with neutron particles inside nuclear reactors.

There are 21 known isotopes of curium, with mass numbers ranging from 232 to 252. Most curium is manufactured in the form of Cm-242 and Cm-244. The isotope with the longest half-life is Cm-247, which takes 16 million years to halve its radioactivity.

Curium is extremely rare and expensive —costing US$4550–5250 per ounce (US$160–185 per mg)—which limits its applications. Cm-242 and Cm-244 are both strong emitters of alpha particles, a weakly penetrating form of radiation (making it easy to contain) but which nevertheless generates a great deal of heat. This has led to uses for curium in so-called radioisotope thermoelectric generators (RTGs), which are incorporated into spacecraft as a long-term power source. Curium has also found an application as an alpha particle source in X-ray spectrometers, used by robotic space probes for establishing the composition and structure of rocks on the Moon and Mars. Curium is fissile, raising the possibility that the metal could be used to build extremely small nuclear weapons.

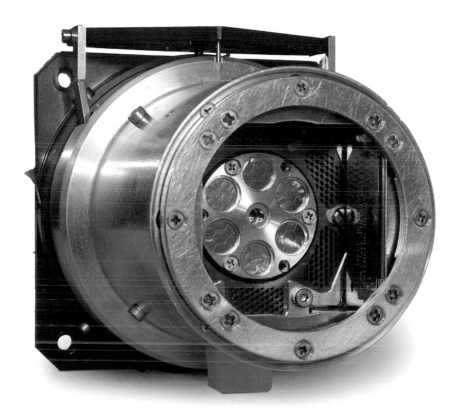

The Alpha Particle X-Ray Spectrometer (APXS) used on NASA's Mars Exploration Rover mission. This device uses small amounts of curium-244 to determine the concentrations of major elements in rocks and soil. It does so by bombarding samples with alpha particles given off by the curium. The paths of the alpha particles as they bounce back, and the energy of X-rays produced as the particles collide with electrons, tell scientists what the sample is made from.

Berkelium

Category: actinide
Atomic number: 97

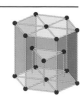

Atomic weight: 247
Color: silver
Phase: solid

Melting point: 1,807 °F (986 °C)
Boiling point: unknown
Crystal structure: hexagonal close-packed

In the 1940s, scientists working as part of the USA's nuclear research program—which had received a massive boost in funding and resources as a result of the Manhattan Project (the program that led to the creation of the atomic bomb)—began to discover a raft of new, heavy chemical elements. These substances weren't uncovered in the normal way of analyzing minerals gathered from the natural world, but rather by cooking up new elements inside nuclear reactors. These new additions to the periodic table became known as the "transuranium" elements—as all were heavier than uranium. In December 1949, the fifth such element was discovered by a team at the University of California at Berkeley. The team called it berkelium.

The choice of name wasn't quite as immodest as it may at first seem. Berkelium is part of the actinide sequence—a string of 15 radioactive chemical elements, running from atomic number 89 to 103. With the actinides left in their correct place, the periodic table becomes cumbersomely wide, so the sequence is appended as a strip at the bottom (under the similar lanthanide sequence). As the actinide elements were discovered, a tradition emerged of naming them "in parallel" with the lanthanide element directly above. For example, the actinide americium sits directly below europium (both named after continents), while curium is underneath gadolinium (both named for famous chemists). Berkelium was directly below terbium, named for the Swedish village of Ytterby, where it was first found, so it seemed only

natural that berkelium be named for its place of origin, too.

The element was manufactured by a team led by pioneering physicist Glenn Seaborg. The team used the Berkeley 60-inch (152-cm) cyclotron to bombard a sample of americium-249 with neutron particles for several hours. They separated the new element using an ion exchange technique and were able to obtain a minute sample. The sample was so minute it was invisible— but was detected chemically later that day. Indeed, this was the easy bit—making enough americium (itself a transuranium element that must be synthesized) to even attempt the experiment had already taken the team five years of laborious work. It would be another nine years before enough of the element could be made that it was visible to the naked eye.

There are 20 known isotopes of berkelium, with mass numbers ranging from 235 to 254. The longest lived of these is Bk-247, with a half-life of 1,400 years, while the least stable has a half-life of just a few microseconds. Very little is known about berkelium, owing to the lack of samples on which to experiment. Even its boiling point—an established property of most elements—is unknown.

Since 1967, the USA's Oak Ridge National Laboratory (ORNL) has managed to produce about 0.035 ounce (1 g) of berkelium, and it costs US$185 per microgram to buy. The one thing it is good for is manufacturing even heavier elements. In 1989 the search for elements 113 to 118 began and was finally completed more than 20 years later.

The first successful cyclotron, built by Ernest
Lawrence at Berkeley, California, in 1930. The
machine has proved extremely important to the
study of the chemical elements. Indeed, element
97, berkelium, was named in honor of the
location where this great device was invented.

Californium

Category: actinide
Atomic number: 98

Atomic weight: 251
Color: silver–white
Phase: solid

Melting point: 1,652 °F (900 °C)
Boiling point: 3,173 °F (1,745 °C)
(estimated)
Crystal structure: double hexagonal

Californium was first produced by a team at the University of California, Berkeley, USA, on February 9, 1950. The discovery was announced in March of that year. The element has the second highest atomic mass of any transuranium element that's been manufactured in quantities large enough to see.

The team, led by Glenn Seaborg, bombarded a sample of curium (page 220) with alpha particles using the university's 60-inch (152-cm) cyclotron—an early form of particle accelerator. Making the californium itself was a fairly quick process; however, making a large enough target of curium to irradiate in the accelerator (just a few millionths of a gram was needed) took them three years. The experiment produced around 5,000 atoms of the californium isotope Cf-245 although, because it has a half-life of just 44 minutes, the sample had decayed away to a single atom just 9.5 hours later.

There are 20 known isotopes of californium. The most stable of these are Cf-251, which has a half-life of 898 years, and Cf-249, with half-life of 351 years. The most common isotope is Cf-252, with a half-life of 2.64 years. Californium is one of the actinides, a sequence of radioactive elements with atomic numbers ranging from 89 to 103. It is a silvery white metal that's very soft and is cut easily with a knife.

With the exception of nuclear test sites, californium is nonexistent in nature due to its extreme radioactivity. That means all californium must be manufactured inside nuclear reactors, usually by bombarding the nuclei of berkelium (page 222) with neutrons. Some of the neutrons are captured by the berkelium nuclei and then decay, by the beta process, into a proton and an electron. The electron is emitted, leaving the proton, which raises the atomic number of the nucleus by one—converting it into a nucleus of californium. The element is generally easier to manufacture than other transuranium elements, a fact that's reflected in its relatively low price, around US$10 per microgram.

The low price is just as well—californium is one of the few transuranium elements to have a number of useful applications. These derive from the copious amount of neutrons given off by the isotope Cf-252, which itself is caused by the tendency of its nuclei to undergo spontaneous fission—splitting apart to release a shower of neutron particles. A microgram of Cf-252 gives off 139 million neutrons per minute.

This has applications in so-called neutron activation analysis, where bombarding a sample of material with neutrons can reveal what elements are hidden inside. The technique can be used to search for precious metals and oil, and to scan baggage for explosives.

A nuclear fuel rod is removed from the High Flux Isotope Reactor (HFIR) at Oak Ridge National Laboratory, USA. The 100-megawatt reactor produces a very high neutron flux and is used to conduct research on manmade elements heavier than plutonium. Its principal product is the synthetic isotope californium-252.

Einsteinium

Category: actinide
Atomic number: 99

Atomic weight: 252
Color: silver
Phase: solid

Melting point: 1,580 °F (860 °C)
Boiling point: unknown
Crystal structure: face-centered cubic

Fittingly, the highly radioactive element einsteinium was discovered in the blast debris from a nuclear bomb test. On November 1, 1952, the USA detonated the "Ivy Mike" nuclear device, in the first successful demonstration of a hydrogen bomb, creating a nuclear explosion based upon the principle of fusion (the bonding together of light atomic nuclei) rather than fission (the splitting apart of heavy ones). Owing to security concerns, the discovery wasn't announced to the world until 1955.

The test was carried out at Enewetak, an atoll in the Pacific Ocean, and produced a blast estimated at just over 10 megatons—equivalent to 10 million tonnes of conventional TNT, and over 700 times as powerful as the Hiroshima bomb (see page 210). Fallout collected from the explosion was found to contain around 200 atoms of the previously unseen element with atomic number 99; a sample large enough to be seen visually was later synthesized in 1961.

The researchers who made the discovery—a team led by Albert Ghiorso, of the University of California, Berkeley, USA—chose to name the element after pioneering German physicist Albert Einstein. It was a wise choice. Einstein's theory of relativity had led to the equation $E=mc^2$, which predicts the vast amount of energy released during a nuclear explosion, as a result of the mass of atoms changing ever so slightly. It also reflects Einstein's political involvement in the creation of the atomic bomb. Einstein and Hungarian physicist Leo Szilard were the first to realize the true threat posed by nuclear weapons, shortly after the discovery of nuclear chain reactions. In 1939, with war looming, the two physicists wrote to President Roosevelt to warn of the potential danger and urge the USA to develop the capability before the Nazis. As it turned out, Germany's nuclear weapons program was a shambles—not helped by the fact that many of the country's best physicists (Einstein included) were Jewish and had thus fled Hitler's regime. Prior to its official christening, the element was jokingly named "pandemonium" after Project PANDA—the research effort that led to "Ivy Mike."

Einsteinium does not occur naturally on Earth and must instead be manufactured inside nuclear reactors. This is done by bombarding plutonium with neutrons, some of which are then absorbed by the plutonium. The irradiated nuclei then decay by beta emission to form the isotope Es-253. This process is capable of manufacturing einsteinium in, typically, milligram quantities; like berkelium (page 222), this scarcity of samples on which to experiment means that there are still unknown properties to einsteinium, including its boiling point.

There are a total of 19 known isotopes of einsteinium, with atomic masses ranging between 240 and 258. The longest lived isotope is Es-252, with a half-life of 470 days. The most commonly occurring isotope is Es-253. It is so radioactive that a gram of it gives off energy at the rate of 1,000 watts – the same as a kettle.

A test tube (approximately 0.35 inch [9 mm] across) in which a tiny sample of einsteinium-253 is undergoing radioactive decay.

Fermium

Category: actinide
Atomic number: 100

Atomic weight: 257
Color: unknown
Phase: solid

Melting point: 2,781 °F (1,527 °C)
Boiling point: unknown
Crystal structure: unknown

Fermium is the 100th element in the periodic table. Like its neighbor einsteinium, it was first detected in the fallout debris from the "Ivy Mike" nuclear test, carried out on the Pacific atoll of Enewetak in 1952. "Ivy Mike" was the first ever thermonuclear explosive: a hydrogen bomb.

The discovery of fermium was made by a team of physicists from the University of California, Berkeley, USA, led by Albert Ghiorso. Particles of fallout had been gathered by aircraft from the mushroom cloud created in the blast, which were then analyzed by the Berkeley team. The new element was formed as neutron particles were captured by nuclei of uranium-238. Neutron capture is generally quite a slow process, but nuclear explosions can generate such a high flux of neutrons (10^{29} per square centimetre per microsecond) that it was possible for a single uranium nucleus to absorb as many as 15 neutrons in one go—followed by eight spontaneous beta decays, converting neutrons into protons and thus raising the atomic number of the nucleus from 92 (uranium) to 100 (fermium).

Fermium is a member of the actinide sequence and, as such, extremely radioactive. For this reason, almost all of the Earth's natural fermium has decayed to nothing, meaning that the element must be synthesized in nuclear reactors. However, the production rate is very small. For example, a flux of 5×10^{15} neutrons per square centimetre per microsecond can be obtained at the USA's Oak Ridge National Laboratory's High Flux Isotope Reactor (much less than that generated in

a nuclear explosion), enabling the reactor to manufacture just picograms of fermium (10^{-12} of a gram) at a time. Annual world production of the element is estimated at about a millionth of a gram. There has never been enough fermium produced to be visible to the naked eye—however, scientists have been able to estimate its properties as being silver in color and reactive to steam, oxygen, and acids. It has no known applications outside of scientific research.

There are 19 known isotopes of fermium, with mass numbers ranging from 242 to 260. The most stable of these is fermium-257, which has a half-life of 100.5 days. Half-lives of the other isotopes range from days down to milliseconds. Fermium-258 has a half-life of just 0.37 milliseconds and this is the principal reason why it's impossible to create any elements with atomic mass numbers heavier than 257 by neutron bombardment alone. For example, if you add a neutron to Fm-257, the resulting nucleus of Fm-258 will have decayed before you've had a chance to add another.

Like einsteinium, fermium is named after one of the pioneers of nuclear physics—in this case, Italian-American scientist Enrico Fermi. In 1942, Fermi lit up the world's first ever nuclear fission reactor, in a disused rackets court at the University of Chicago, USA. Fermi stacked up a pile of uranium and graphite blocks (to slow the neutrons). Neutron-absorbing cadmium rods were inserted to give Fermi some control over the reaction. Another set of rods hung above the reactor on ropes—to be dropped into the core and shut down the reaction in an emergency.

A particle deflection magnet on display in Fermilab's Wilson Hall, Illinois, USA. Magnets like this are used to guide beams of subatomic particles inside a particle accelerator. Both Fermilab and the chemical element fermium are named in honor of the pioneering physicist Enrico Fermi.

The Transfermium Elements

There came a point in the discovery of the chemical elements when scientists stopped looking for traces of hitherto unknown substances in terrestrial rocks and minerals and instead began manufacturing new elements in their laboratories. Many of the elements in the actinide sequence—the block of elements spanning atomic numbers 89 to 103—were discovered this way. Indeed, this was the case for all of the elements above uranium, atomic number 92. Elements 93 (neptunium) to 100 (fermium) were detailed on pages 214–229. Here we look at the group beyond fermium, with atomic numbers above 100—the so-called "transfermium" elements.

As we move up through the periodic table, elements have increasing numbers of positively charged proton particles in their nuclei. To stop the nuclei flying apart by electrostatic repulsion, the proton particles have to be separated by large numbers of electrically neutral neutron particles. Because neutrons and protons each weigh one atomic mass unit, this makes the nuclei of the transfermium elements extremely heavy. For example, plutonium has 94 protons in its nucleus, and a common isotope has 144 neutrons, giving a total of 238 atomic mass units—each atom weighs 238 times the mass of a hydrogen atom.

They are also highly radioactive. The mass of an atomic nucleus increases at the expense of its stability. Imagine the nucleus as rather like a huge, quivering drop of water—the bigger it gets, the more likely it is to spontaneously break apart into smaller droplets. Atoms don't break into droplets, but they do decay—through radioactivity—and this is why atoms with heavy atomic nuclei tend to be the most radioactive. This means that all of the transfermium elements that were present on Earth when it formed have now decayed away completely. In order for these elements to be discovered, they have to be manufactured in the laboratory.

Perhaps the simplest way to do this is by adding enough neutrons to an existing atomic nucleus, so that it becomes unstable and decays into a new element. This might sound like a contradiction, but in fact there are modes of radioactive "decay" that enable elements to transmute into new elements with a higher atomic number. One such process is known as "beta-minus" decay, in which a neutron decays to become a proton, plus an electron, and a ghostly subatomic particle called a neutrino. Neutrinos have no mass or charge, just a strange, quantum mechanical property called "spin." Both the electron and the neutrino escape the nucleus as "radiation," while the proton remains. The net effect is to raise the atomic number by one—thus creating a nucleus of the next element in the periodic table.

Beta-minus decay is one of several forms of radioactivity. There's also "beta-plus" decay, where a proton turns into a neutron and a positron (the antimatter counterpart of the electron) along with a neutrino. Another possibility is alpha decay. An "alpha particle" is essentially a helium nucleus—a cluster of two neutrons and two protons—that can be spontaneously spat out by a radioactive nucleus. Similarly, some nuclei, when collided with an alpha particle, will absorb the particle, instantly raising their atomic number by two—creating an element two places higher on the periodic table.

The radioactivity of an element is quantified by a number known as the half-life. Highly radioactive elements have short half-lives and decay quickly; conversely, more stable elements have long half-lives and decay slowly. Radioactive decay is a random process—each element decays

The Transfermium Elements

Name	Chemical symbol	Atomic number	Category	Group	Date of discovery	Place of discovery	Named for...
Mendelevium	Md	101	Actinide	n/a	1955	UC Berkeley (USA)	Dmitri Mendeleev (Russian chemist)
Nobelium	No	102	Actinide	n/a	1966	JINR (Rus)	Alfred Nobel (Swedish chemist)
Lawrencium	Lr	103	Actinide	n/a	1961	UC Berkeley (USA)	Ernest Lawrence (American physicist)
Rutherfordium	Rf	104	Transition metal	4	1966/69	JINR (Rus)/ UC Berkeley (USA)	Ernest Rutherford (New Zealand/British physicist)
Dubnium	Db	105	Transition metal	5	1968/70	JINR (Rus)/ UC Berkeley (USA)	Dubna (Russian town)
Seaborgium	Sg	106	Transition metal	6	1974	LBNL/LLNL (USA)	Glenn Seaborg (American chemist)
Bohrium	Bh	107	Transition metal	7	1981	GSI (Ger)	Niels Bohr (Danish physicist)
Hassium	Hs	108	Transition metal	8	1984	GSI (Ger)	Hesse (German state)
Meitnerium	Mt	109	unknown	9	1982	GSI (Ger)	Lise Meitner (Austrian physicist)
Darmstadtium	Ds	110	unknown	10	1994	GSI (Ger)	Darmstadt (German city)
Roentgenium	Rg	111	unknown	11	1994	GSI (Ger)	Wilhelm Konrad Roentgen (German physicist)
Copernicium	Cn	112	Transition metal	12	1996	GSI (Ger)	Nicolaus Copernicus (Polish astronomer)
Ununtrium	Uut	113	unknown	13	2003 (claimed)	TBC	n/a
Flerovium	Fl	114	unknown	14	1999	JINR (Rus)	Georgy Flyorov (Russian physicist)
Ununpentium	Uup	115	unknown	15	2003 (claimed)	TBC	n/a
Livermorium	Uuh	116	unknown	16	2000	JINR (Rus)	Livermore (Californian town)
Ununseptium	Uus	117	unknown	17	2010 (claimed)	TBC	n/a
Ununoctium	Uuo	118	unknown	18	2002 (claimed)	TBC	n/a

Dmitri Mendeleev was initially a mediocre student, but left college at the top of his class. He later became a scientist, classifying and organizing the chemical elements by their weights and properties. From this work, he developed the first true periodic table of the elements, which was published in 1869.

spontaneously but in accordance with well-defined statistical rules. The half-life is a statistical property that quantifies the average time it takes for half of the atoms in a sample to have randomly decayed into other elements. For example, 64 atoms of an element that has a half-life of a second will on average (remember, it's a statistical process) have decayed to 32 atoms after 1 second, 16 after 2 seconds, 8 after 3, 4 after 2, then 2 and 1.

The transfermium elements have been discovered by researchers working, principally, in three laboratories around the world: the University of California, Berkeley, in the USA; the Joint Institute for Nuclear Research (JINR), in Dubna, Russia; and the Center for Heavy Ion Research in Darmstadt, Germany. Sometimes, owing in part to the short half-lives of these elements, coupled with the sheer difficulty in manufacturing

them, only a few atoms of some elements have ever been made.

In the short term, new elements are assigned temporary names according to a well-defined set of rules—until the discovery has been confirmed and a permanent name agreed upon. In this naming scheme, each of the three digits in the element's atomic number corresponds to a syllable, according to the rule: 0 = nil, 1 = un, 2 = bi, 3 = tri, 4 = quad, 5 = pent, 6 = hex, 7 = sept, 8 = oct, and 9 = en. The element's name is then finished with the three letters "ium." So, for example, element 118 is "ununoctium."

Deciding on a permanent name is a somewhat more complex process. Centuries and decades ago, new elements were discovered regularly. However, new discoveries are now few and far between and when one does come along the battle for credit and—crucially—the right to name it, is fierce. In 1985, a considerable spat erupted between Berkeley and the JINR over the naming of elements 104–107, with both institutes claiming credit. The dispute bordered on childishness, with the American Chemical Society at one point ordering its research journals not to use any of the proposed Russian nomenclature for these elements. The matter was finally resolved in 1997—but only after years of negotiation.

Elements have now been confirmed, and given permanent names, up to atomic number 112, plus 114 (flevorium) and 116 (livermorium) were given full status in 2012. There have been discoveries claimed for elements 113, 115, 117, and 118, but they await ratification by the International Union of Pure and Applied Chemistry. Element 118 neatly completes period 7 of the periodic table. At the time of writing no elements have been found beyond this and so there is, as yet, no need to add an extra row to the table in the form of period 8. Although it's unlikely that this will remain the case for ever.

Glossary

Acid—Acids are compounds that contain hydrogen. When mixed with water, the hydrogen is released as positive ions that can bond with other substances, having a corrosive effect on them.

Actinides—A sequence of highly radioactive chemical elements spanning atomic numbers 89 to 103.

Alkali metal—A category of elements containing most of group 1 of the periodic table: specifically, lithium, sodium, potassium, rubidium, cesium, and francium. They are all soft, shiny, highly reactive metals.

Alkaline earth metal A category of reactive metal elements comprising all the periodic table's group 2: beryllium, magnesium, calcium, strontium, barium, and radium.

Allotropes—These are forms of the same element that have their atoms arranged in different ways. For example, soot and diamond are different allotropes of carbon.

Alloy—A mixture of two or more metal elements, often produced to give desirable engineering properties.

Alpha radiation—A kind of radioactive decay particle, comprising two neutrons and two protons in a tight cluster.

Amorphous solid—The opposite of a crystal, an amorphous solid is a material in which atoms are arranged with no regular structure. An example is glass.

Atom—The fundamental unit of a chemical element. An atom is made of protons and neutrons concentrated into a central nucleus, around which electrons orbit. Unless the atom is "ionized" (see below), there will be the same number of protons as electrons.

Atomic mass number—A measure of the mass of an atom, given by adding up the total number of neutrons and protons contained in its nucleus.

Atomic number—A measure of the charge on an atom's nucleus, given by adding up the total number of protons it contains. A chemical element can be uniquely identified by determining the atomic number of its atoms.

Atomic weight—The actual mass of an atom measured in units equal to 1/12 the mass of the isotope carbon-12. When different isotopes of an element exist in nature, the atomic weight is normally an average weighted by each isotope's abundance.

Avogadro number—The number of particles in one "mole" of a substance, equal to $6.02214199 \times 10^{23}$. The number is defined such that a mole of carbon-12 atoms weighs exactly 12 grams.

Avogadro's law—A law which says that equal volumes of gas, at equal temperature and pressure, contain an equal number of particles.

Base—A base, sometimes called an "alkali," is the opposite of an acid. It soaks up the positive ions to reduce the acidity of a solution.

Beta radiation—A process by which protons and neutrons can convert into one another, emitting either an electron or its antimatter counterpart, the positron.

Catalyst—A substance that increases the rate of a chemical reaction without being involved in the reaction itself. For example, in some cars catalytic converters use a platinum catalyst to encourage poisonous carbon monoxide in the car's exhaust to combine with oxygen in the air to form carbon dioxide.

Chemical reaction—When two different elements or compounds come together, a chemical reaction may result as electrons in outer layers of the atoms and molecules interact with one another. A chemical reaction can produce heat and light, changes in color and the formation of new chemical compounds.

Compound—Atoms of elements bond together to form a new substance.

Conductivity—Both heat and electricity can be conducted through a material by free electrons, which slosh around inside it, not tied to any particular atoms. Because metals have more free electrons, they are generally better conductors than nonmetals.

Conservation of mass—During a chemical reaction, there is no appreciable change in the total mass of the reactants. This is the law of conservation of mass. Mass is not conserved in nuclear reactions, when a large amount can be converted into energy.

Cosmic nucleosynthesis— All of the chemical elements in the universe were either created from fundamental subatomic particles in the Big Bang in which our universe was born (which made mostly hydrogen and helium), forged in nuclear reactions taking place inside the hot cores of stars (which made pretty much everything else), or they were produced by the radioactive decay of such elements.

Crystal structure—The arrangement of atoms inside a solid chemical element. Cubic and hexagonal are among the many possibilities.

Cyclotron—An early design of particle accelerator in which charged bodies, such as protons and alpha particles, are accelerated to high speed by electric and magnetic fields.

Electron—A low-mass, negatively charged variety of subatomic particle, which inhabits the outer layers of atoms. Interactions between electrons in different atoms and molecules are what cause chemical reactions.

Electron shell—Electrons are organized into layers within atoms known as "shells," the structure of which are determined by quantum theory. The number of electrons in each electron shell is what determines the chemical properties of an element.

Element—An element is a chemical substance whose fundamental units are atoms. More complex chemical compounds are composed of molecules, which consist of atoms bonded together.

Element symbol—A two-letter abbreviation for the name of a chemical element. Usually, the letters in the symbol are derived from the element's name. Sometimes they derive from an ancient name for the element. For example, Pb is the symbol for lead, which comes from its Latin name, *plumbum*.

Fission—A nuclear reaction in which a heavy atomic nucleus splits in half to form two lighter nuclei, releasing energy in the process. This is how early nuclear weapons (such as the Hiroshima bomb) worked and is still the principle underpinning modern nuclear power stations.

Fusion—A nuclear reaction in which two light atomic nuclei fuse together to make a single heavier one, giving off energy in the process. This is the process that powers the Sun and other stars, and is the basis for modern "thermonuclear" weapons.

Group—Vertical columns in the periodic table are called groups. The elements in a single group all have a similar number of particles in the valence shell (see page 236), giving them all very similar chemical properties.

Half-life—Radioactive chemical elements are characterized by their half-life. This is the time it takes for half the atoms in a sample of the element to have undergone radioactive decay.

Radioactive elements have short half-lives; stable elements have long half-lives.

Halogen—"The halogens" are group 17 of the periodic table, containing the nonmetal, often gaseous elements fluorine, chlorine, bromine, iodine, and astatine.

Heavy water—Water has the chemical formula H_2O—two hydrogen atoms and an oxygen atom. If the hydrogen is replaced by deuterium (an isotope of hydrogen with an extra neutron in the nucleus), then the result is D_2O, or heavy water.

Helium-3—Helium-3 is an isotope of helium that has one less neutron in its nucleus. Found in abundance on the Moon, it's desirable as a fuel for future nuclear fusion power stations.

Ideal gas—An ideal gas is a theoretical model in physics for how gases behave. It treats the molecules in the gas as completely dimensionless points, and ignores the effects of intermolecular forces.

Intermolecular force— Forces that operate between molecules of a gas. They are caused by the electrical charge of the electrons (negative) and protons (positive) that the molecules are made of.

Ion—Normally, an atom has the same number of positively charged protons and negatively charged electrons, giving it zero net charge. An atom that has had electrons either added or removed to give it nonzero charge is known as an ion. Sometimes ions are subdivided into "anions" (negatively charged) and "cations" (positively charged).

Ion exchange—A way of isolating elements from other chemicals by extracting their electrically charged ions from a solution.

Isotope—Different isotopes of a chemical element contain different numbers of neutrons in their nuclei. The most common natural isotope of carbon, for example, has six neutrons with its six protons, giving a total atomic mass number of 12, usually written "carbon-12." Other isotopes are carbon-13 and carbon-14, with seven and eight neutrons respectively.

IUPAC—The International Union of Pure and Applied Chemistry, the organization that sets international standards for the science of chemistry and approves the names of newly discovered chemical elements.

Lanthanides—Also known as the "rare earth" elements, the lanthanide sequence runs from atomic number 57 to 71. It normally appears appended at the foot of the periodic table.

Laser—A laser (Light Amplification by the Stimulated Emission of Radiation) is a quantum mechanical device that generates "coherent" light —that is, light waves of the same wavelength, traveling in the same direction.

Metalloid—The metalloids are a diagonal category of elements, with properties between those of metals and nonmetals. They include

arsenic, germanium, and silicon and are often used in semiconductor electronics.

Mixture—When atoms or molecules are combined but do not bond with one another chemically, the resulting substance is known as a mixture—in contrast to a compound, in which chemical bonding does take place.

Molecule—When two or more atoms bond together, the result is called a molecule. Molecules are the fundamental units of compounds and have a very rich spectrum of chemical properties.

Neutrino—A subatomic particle that has neither mass nor electric charge. Neutrinos are an essential component of particle physics, however, because they carry an important quantum mechanical property called "spin."

Neutron—A neutron is a subatomic particle with about the same mass as a proton, but zero electrical charge.

Noble gas—The elements in the right-hand column of the periodic table (group 18) are known collectively as the noble gases, or sometimes the "inert" gases because they are unreactive. They include neon, argon, and xenon.

Nonmetals—Nonmetal is a classification concentrated around the upper-right of the periodic table (which contains carbon, oxygen as well as sulfur), but also includes the top-left element of the table, hydrogen.

Nucleus—An atom's core,

consisting of a cluster of protons and neutrons, bound together by a quantum mechanical phenomenon known as the "strong force."

Nuclear reactor—A device for the creation of nuclear reactions to alter the nucleus of an atom. This can be done to generate energy or to manufacture rare elements— using nuclear processes to change the "atomic number" of elements.

Ore—A rock that contains minerals, which are a kind of chemical compound. By crushing up ore and subjecting the mineral compounds to chemical processes, it's possible to extract the pure chemical elements from which the minerals are made.

Oxide—An oxide is a chemical compound containing oxygen. Oxygen in the air reacts with almost every element to form an oxide. A common example is iron oxide (rust).

Period—Each row of the periodic table is known as a "period," reflecting the fact that the properties of the elements repeat down the table.

Pitchblende—A radioactive, uranium-rich ore, containing uranium dioxide, as well as compounds of lead, thorium, and other elements.

Phases of matter—Matter comes in three principal phases: solid, liquid, and gas. Substances can exist in all of these states depending on their temperature and pressure. At standard atmospheric pressure for example, water is solid below 0 °C, liquid between 0 and 100 °C, and is a gas above 100 °C.

Post-transition metal—The post-transition metals are found to the right of the transition metals in the periodic table. They are soft metals with low melting and boiling points.

Pressure—This is force per unit area. A confined gas or liquid has an inherent pressure inside it. For example, at the surface of the Earth our atmosphere exerts a pressure of 14.7 pounds per square inch (1.03 kg per square cm). Pressure is measured in Pascals (Pa)—1 Pa is equal to 1 kg (in Earth gravity) per square meter.

Proton—A positively charged subatomic particle with about the same mass as the neutron.

Quantum theory—A branch of physics governing the behavior of subatomic particles and radiation, derived from the idea that energy cannot take a continuous range of values but instead comes in discrete, indivisible packages, or "quanta."

Radioactivity—The spontaneous disintegration of the nucleus of an atom. Heavy atomic nuclei are typically unstable, due to their size, making them radioactive. The particles and radiation given off by highly radioactive nuclei are hazardous to health.

RTGs—Radioisotope thermoelectric generators are devices for generating electricity on deep-space probes. They work by harnessing the heat generated by the decay of radioactive elements such as plutonium.

Semiconductor—A material with electrical conductivity between that of a metal and a nonmetal. Examples include silicon and germanium.

Spectroscopy—An analytical method for determining the presence of particular chemical elements in a sample. It often involves burning the sample and looking for the presence of particular "lines"—peaks and troughs seen when the brightness of the light given off is plotted against its wavelength. The lines are uniquely determined by the electron shell structure of the atoms producing them.

STP—Standard temperature and pressure. When discussing the phase of a substance (solid, liquid, or gas) it is usual to quote the phase at STP—temperature zero °C and pressure 100,000 Pascals.

Superconductor—A material with zero electrical resistance. It is a perfect conductor of electricity. Normally, superconductors have to be cooled close to absolute zero (-273.15 °C/-459.6 °F) to operate, but the search is on to find chemical compounds that become superconducting at higher temperatures.

Synthetic element—Some elements are so radioactive that they have long decayed away on Earth. The only way to see such elements is to make them in the laboratory. These are known as the synthetic elements.

Temperature—A physical property determining how hot or cold an object is. Physicists ascribe temperature to vibrations of the atoms or molecules in a substance, with high temperature corresponding to vigorous vibrations.

Trace element—A trace element is defined as a chemical element that has a concentration of less than 100 parts per million.

Transuranium elements—Elements with atomic numbers greater than 92 (the atomic number of uranium). The atomic nuclei of these elements are so big that they are all unstable, and therefore radioactive.

Transition metal—The transition metals make up the bulk of the periodic table, spanning most of groups 3–12.

Valence shell—The outermost electron shell of an atom is known as the "valence" shell. It plays the dominant role in chemical reactions. Chemical elements occupying the same "group" in the periodic table have the similar number of electrons in their valence shell.

Index

THE AUTHORS

Dr Paul Parsons is a science journalist and author based in southeast England. He was formerly editor of the BBC's award-winning science and technology magazine *Focus*. His book *The Science of Doctor Who* was longlisted for the 2007 Royal Society Prize for Science Books.

Gail Dixon is a journalist and editor who has worked on a variety of magazines, including *Focus*, one of Britain's leading science journals. She recently coauthored *3-minute Hawking*, the most up-to-date book on the greatest living theoretical physicist.

CREDITS

New York • London

© 2013 by Quercus Editions Ltd
First published in the United States by Quercus in 2014

Any member of educational institutions wishing to photocopy part or all of the work for classroom use or anthology should send inquiries to Permissions c/o Quercus Publishing Inc., 31 West 57th Street, 6th Floor, New York, NY 10019, or to permissions@quercus.com.

ISBN 978-1-62365-110-7

Library of Congress Control Number: 2013913390

Distributed in the United States and Canada by Random House Publisher Services
c/o Random House, 1745 Broadway
New York, NY 10019

Manufactured in China

2 4 6 8 10 9 7 5 3 1

www.quercus.com

PICTURE CREDITS

p4 © Charles D. Winters / Science Photo Library; p11 © Charles D. Winters / Science Photo Library; p12 © US Department Of Energy / Science Photo Library; p15 © Science Photo Library; p17 © Science Photo Library; p19 © Science Photo Library; p21 © Science Photo Library; p23 © Science Photo Library; p25 © Charles D. Winters / Science Photo Library; p27 © Tatjana Romanova / Shutterstock; p28 © Susumu Nishinaga / Science Photo Library; p31 © Michael Clutson / Science Photo Library; p33 © Science Photo Library; p35 © Science Photo Library; p37 © Science Photo Library; p39 © Science Photo Library; p41 © Science Photo Library; p42 © Astrid & Hans Frieder Michler / Science Photo Library; p45 © Science Photo Library; p47 © Natural History Museum, London / Science Photo Library; p49 © Science Photo Library; p51 © Science Photo Library; p53 © Andrew Lambert Photography / Science Photo Library; p55 © Science Photo Library; p57 © Science Photo Library; p59 and back cover © Science Photo Library; p61 © Science Photo Library; p63 © Science Photo Library; p65 © Science Photo Library; p67 © Science Photo Library; p68 © Steve Bloom Images/Alamy; p71 © Science Photo Library; p73 © Science Photo Library; p75 © Scientifica /Visuals Unlimited, Inc. / Science Photo Library; p77 © Science Photo Library; p79 © Science Photo Library; p81 © Science Photo Library; p83 © Natural History Museum, London / Science Photo Library; p85 © Science Photo Library; p87 © Andrew Lambert Photography / Science Photo Library; p89 © Science Photo Library; p90 © Charles O'Rear/Corbis; p93 © Science Photo Library; p95 © Science Photo Library; p97 © Science Photo Library; p99 © Science Photo Library; p101 © Science Photo Library; p103 © Science Photo Library; p105 © Scott Camazine / Science Photo Library; p107 © Science Photo Library; p109 © Science Photo Library; p111 © Science Photo Library; p113 © Science Photo Library; p114 © Eye Of Science / Science Photo Library; p117 © Science Photo Library; p119 © GiPhotostock / Science Photo Library; p121 © Science Photo Library; p123 © Science Photo Library; p125 © Science Photo Library; p127 © Claude Nuridsany & Marie Perennou / Science Photo Library;

p129 © Rich Treptow / Science Photo Library; p131 © Science Photo Library; p133 © Gary Cook/Visuals Unlimited, Inc. / Science Photo Library; p135 © Science Photo Library; p136 © Eye Of Science / Science Photo Library; p139 © Science Photo Library; p141 © Science Photo Library; p143 © Science Photo Library; p145 © Theodore Gray/Visuals Unlimited/Corbis; p147 © Science Photo Library; p149 © Science Photo Library; p151 © Science Photo Library; p153 © Science Photo Library; p155 © Science Photo Library; p157 © Science Photo Library; p159 © Science Photo Library; p161 © Science Photo Library; p163 © Science Photo Library; p165 © Science Photo Library; p167 © Science Photo Library; p169 © Science Photo Library; p171 © Science Photo Library; p172 © Chris Knapton / Science Photo Library; p175 © Science Photo Library; p177 © Science Photo Library; p179 © Science Photo Library; p181 © Ria Novosti / Science Photo Library; p183 © David Nunuk / Science Photo Library; p184 © Duncan Smith/Corbis; p187 © ImageState/Alamy; p189 © Charles D. Winters / Photo Researchers, Inc. / Science Photo Library; p191 © Science Photo Library; p193 © Science Photo Library; p195 © Astrid & Hanns-Frieder Michler / Science Photo Library; p197 © Theodore Gray/Visuals Unlimited/Corbis; p199 © Health Protection Agency / Science Photo Library; p201 © Dirk Wiersma / Science Photo Library; p203 © J.C. Revy, ISM / Science Photo Library; p205 © Ted Kinsman / Photo Researchers, Inc. / Science Photo Library; p207 © Science Photo Library; p209 © Dirk Wiersma / Science Photo Library; p211 © Dirk Wiersma / Science Photo Library; p212 © US Department Of Energy / Science Photo Library; p215 © Dirk Wiersma / Science Photo Library; p217 © Los Alamos National Laboratory / Science Photo Library; p219 © Bionerd; p221 © NASA; p225 © US Department Of Energy / Science Photo Library; p227 © Haire, R. G. / US Department of Energy; p229 © Mark Williamson / Science Photo Library; p231 © Volker Steger / Science Photo Library; p232 © Ria Novosti / Science Photo Library.

All illustrations by Mark Franklin.